STECK-VAUGHN

W9-CFZ-747

BASIC ESSENTIALS OF
MATHEMATICS

BOOK ONE

WHOLE NUMBERS, FRACTIONS, & DECIMALS

AUTHOR
James T. Shea

CONSULTANT
Brenda C. Ramsey
ABE/GED Instructor
Memphis City Schools

ACKNOWLEDGMENTS

Senior Math Editor: Karen Lassiter, Ph.D.
Project Coordinator: Joyce Spicer

Project Design and Development:
The Wheetley Company

Cover Design and Illustration:
James Masch

STECK-VAUGHN ADULT EDUCATION ADVISORY COUNCIL

Donna D. Amstutz
Assistant Professor
Northern Illinois University
DeKalb, Illinois

Sharon K. Darling
President, National Center
  for Family Literacy
Louisville, Kentucky

Roberta Pittman
Director, Project C3 Adult Basic Education
Detroit Public Schools
Detroit, Michigan

Elaine Shelton
President, Shelton Associates
Consultant, Competency-Based Adult Education
Austin, Texas

STECK-VAUGHN
C O M P A N Y
ELEMENTARY · SECONDARY · ADULT · LIBRARY

ISBN: 0-8114-4668-9
Copyright © 1991 Steck-Vaughn Company.
Printed in the United States of America.

Table of Contents

To the Student

The *Basic Essentials of Mathematics* is a two-book basic math program that teaches whole number, fraction, and decimal skills in Book 1 and teaches percent, measurement, formulas, equations, ratio, and proportion skills in Book 2. Both books have been designed and written for the adult learner who wants to brush up on math skills in a minimum amount of time. These books allow adult learners to act as their own teacher and check their own work. This way, each learner can progress at his or her own rate.

Each page is a separate lesson that deals with a single skill. Many pages contain a box at the top where example problems are worked out in a step-by-step procedure. The remainder of the page contains practice problems similar to the ones worked out in the example box at the top of the page. Frequently, the first practice problem is worked out, demonstrating how to arrive at the correct answer.

The Answer Key on page 119 allows students to check their own work. Checking Up pages and Unit Review pages provide frequent review of previously taught skills. These review pages can be used to indicate mastery of the skills taught or the need for additional practice.

To help develop skills in solving problems, three pages of problem-solving strategies appear in each book. These strategies include Choose an Operation, Find a Pattern, Use Estimation, Multi-step Problems, Make a Diagram, and Identify Extra Information. Other lessons develop skills in estimating and rounding numbers.

Each of the two books concludes with a Mastery Test of 100 items. These tests are divided into skill areas. A score chart is included. If a student misses more than two problems in any skill area, he or she should go back and review that skill. When Book 1 is mastered, students may proceed with confidence to Book 2.

Unit 1 Whole Numbers
Basic Addition Facts

Study and master these addition facts.

$$\begin{array}{r} 3 \leftarrow \text{addend} \\ +\,3 \leftarrow \text{addend} \\ \hline 6 \leftarrow \text{sum} \end{array}$$

Find the sums.

1.
$\begin{array}{r}0\\+1\\\hline 1\end{array}$ $\begin{array}{r}2\\+1\\\hline 3\end{array}$ $\begin{array}{r}3\\+1\\\hline 4\end{array}$ $\begin{array}{r}4\\+1\\\hline 5\end{array}$ $\begin{array}{r}5\\+1\\\hline 6\end{array}$ $\begin{array}{r}6\\+1\\\hline 7\end{array}$ $\begin{array}{r}7\\+1\\\hline 8\end{array}$ $\begin{array}{r}8\\+1\\\hline 9\end{array}$ $\begin{array}{r}9\\+1\\\hline 10\end{array}$

2.
$\begin{array}{r}1\\+2\\\hline 3\end{array}$ $\begin{array}{r}5\\+2\\\hline 7\end{array}$ $\begin{array}{r}2\\+2\\\hline 4\end{array}$ $\begin{array}{r}9\\+2\\\hline 11\end{array}$ $\begin{array}{r}7\\+2\\\hline 9\end{array}$ $\begin{array}{r}8\\+2\\\hline 10\end{array}$ $\begin{array}{r}0\\+2\\\hline 2\end{array}$ $\begin{array}{r}6\\+2\\\hline 8\end{array}$ $\begin{array}{r}4\\+2\\\hline 6\end{array}$

3.
$\begin{array}{r}0\\+3\\\hline 3\end{array}$ $\begin{array}{r}3\\+3\\\hline 6\end{array}$ $\begin{array}{r}5\\+3\\\hline 8\end{array}$ $\begin{array}{r}7\\+3\\\hline 10\end{array}$ $\begin{array}{r}9\\+3\\\hline 12\end{array}$ $\begin{array}{r}2\\+3\\\hline 5\end{array}$ $\begin{array}{r}4\\+3\\\hline 7\end{array}$ $\begin{array}{r}6\\+3\\\hline 10\end{array}$ $\begin{array}{r}8\\+3\\\hline 11\end{array}$

4.
$\begin{array}{r}2\\+4\\\hline 6\end{array}$ $\begin{array}{r}0\\+4\\\hline 4\end{array}$ $\begin{array}{r}6\\+4\\\hline 9\end{array}$ $\begin{array}{r}8\\+4\\\hline 12\end{array}$ $\begin{array}{r}1\\+4\\\hline 5\end{array}$ $\begin{array}{r}3\\+4\\\hline 7\end{array}$ $\begin{array}{r}5\\+4\\\hline 9\end{array}$ $\begin{array}{r}7\\+4\\\hline 11\end{array}$ $\begin{array}{r}9\\+4\\\hline 13\end{array}$

5.
$\begin{array}{r}1\\+5\\\hline 6\end{array}$ $\begin{array}{r}3\\+5\\\hline 8\end{array}$ $\begin{array}{r}5\\+5\\\hline 10\end{array}$ $\begin{array}{r}7\\+5\\\hline 12\end{array}$ $\begin{array}{r}9\\+5\\\hline 14\end{array}$ $\begin{array}{r}0\\+5\\\hline 5\end{array}$ $\begin{array}{r}4\\+5\\\hline 9\end{array}$ $\begin{array}{r}6\\+5\\\hline 11\end{array}$ $\begin{array}{r}8\\+5\\\hline 13\end{array}$

6.
$\begin{array}{r}1\\+6\\\hline 7\end{array}$ $\begin{array}{r}3\\+6\\\hline \end{array}$ $\begin{array}{r}2\\+6\\\hline \end{array}$ $\begin{array}{r}4\\+6\\\hline \end{array}$ $\begin{array}{r}5\\+6\\\hline \end{array}$ $\begin{array}{r}7\\+6\\\hline \end{array}$ $\begin{array}{r}6\\+6\\\hline \end{array}$ $\begin{array}{r}0\\+6\\\hline \end{array}$ $\begin{array}{r}9\\+6\\\hline \end{array}$

7.
$\begin{array}{r}1\\+7\\\hline 8\end{array}$ $\begin{array}{r}4\\+7\\\hline 11\end{array}$ $\begin{array}{r}7\\+7\\\hline 14\end{array}$ $\begin{array}{r}0\\+7\\\hline 7\end{array}$ $\begin{array}{r}5\\+7\\\hline 12\end{array}$ $\begin{array}{r}8\\+7\\\hline 15\end{array}$ $\begin{array}{r}3\\+7\\\hline 10\end{array}$ $\begin{array}{r}6\\+7\\\hline 13\end{array}$ $\begin{array}{r}9\\+7\\\hline 16\end{array}$

8.
$\begin{array}{r}0\\+8\\\hline 8\end{array}$ $\begin{array}{r}5\\+8\\\hline 13\end{array}$ $\begin{array}{r}9\\+8\\\hline 17\end{array}$ $\begin{array}{r}2\\+8\\\hline 10\end{array}$ $\begin{array}{r}6\\+8\\\hline 14\end{array}$ $\begin{array}{r}3\\+8\\\hline 11\end{array}$ $\begin{array}{r}7\\+8\\\hline 15\end{array}$ $\begin{array}{r}8\\+8\\\hline 16\end{array}$ $\begin{array}{r}4\\+8\\\hline 12\end{array}$

9.
$\begin{array}{r}1\\+9\\\hline 10\end{array}$ $\begin{array}{r}6\\+9\\\hline 15\end{array}$ $\begin{array}{r}2\\+9\\\hline 11\end{array}$ $\begin{array}{r}7\\+9\\\hline 16\end{array}$ $\begin{array}{r}3\\+9\\\hline 12\end{array}$ $\begin{array}{r}8\\+9\\\hline 17\end{array}$ $\begin{array}{r}4\\+9\\\hline 13\end{array}$ $\begin{array}{r}9\\+9\\\hline 18\end{array}$ $\begin{array}{r}5\\+9\\\hline 14\end{array}$

Reading and Writing Whole Numbers

A place-value chart can help you understand whole numbers. Each digit in a number has a value based on its place in the number.

The 7 is in the millions place.
Its value is 7 millions or 7,000,000.

The 1 is in the ten thousands place.
Its value is 1 ten thousand or 10,000.

The 4 is in the hundreds place.
Its value is 4 hundreds or 400.

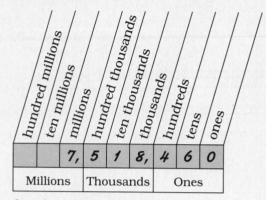

We read and write this number as: seven million, five hundred eighteen thousand, four hundred sixty.

Notice that commas are used to separate digits into groups of three. This helps make large numbers easier to read.

Complete the following exercises.

1. Write 1,954 in words _____ *one* _____ thousand, _____ *nine* _____ hundred _____ *fifty-four* _____ .

2. 1,917 is written _____ thousand, _____ hundred _____ .

3. 1,812 is written _____ thousand, _____ hundred _____ .

4. 25,416 is written _____ thousand, _____ hundred _____ .

5. 14,703 is written _____ thousand, _____ hundred _____ .

6. 10,908 is written _____ thousand, _____ hundred _____ .

7. 12,100 is written _____ thousand, _____ hundred.

8. 12,008 is _____ thousand, _____ .

9. 10,092 is _____ thousand, _____ .

10. 123,456 is _____ hundred _____ thousand, _____ hundred _____ .

11. 756,100 is _____ hundred _____ thousand, _____ hundred.

12. 1,658,325 is _____ million, _____ hundred _____ thousand, _____ hundred _____ .

Rounding

Rounded numbers tell *about* how many. Use a number line to help you round numbers. When a number is halfway between numbers, always round up to the larger number.

Round 43 to the nearest ten.

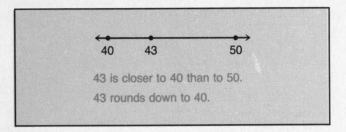

43 is closer to 40 than to 50.

43 rounds down to 40.

Round 750 to the nearest hundred.

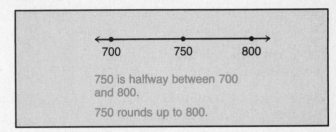

750 is halfway between 700 and 800.

750 rounds up to 800.

Here is another way to round.

Round 37,472 to the nearest thousand.

When the digit to the immediate right of the digit you want to round is **4 or less,** round down and replace all digits to the right with zeros.

37,472 rounds down to 37,000. (since 4 is **4 or less**)

Round 38,673 to the nearest hundred.

When the digit to the immediate right of the digit you want to round is **5 or more,** round up and increase the desired digit's value by one. Replace the digits to the right of this digit with zeros.

38,673 rounds up to 38,700. (since 7 is **5 or more**)

Round to the nearest ten.

1. 57 ___60___   82 _____   49 _____   35 _____   51 _____

2. 743 ___740___   4,765 _____   3,644 _____   5,555 _____   49,999 _____

Round to the nearest hundred.

3. 782 ___800___   456 _____   327 _____   218 _____

4. 68,496 ___68,500___   56,555 _____   24,098 _____   79,502 _____

Round to the nearest thousand.

5. 2,537 ___3,000___   5,499 _____   6,205 _____   3,668 _____

6. 18,305 ___18,000___   44,174 _____   36,745 _____   11,831 _____

Write these numbers as round numbers.

EXACT NUMBER	TO NEAREST TEN	TO NEAREST HUNDRED	TO NEAREST THOUSAND	TO NEAREST MILLION
7. 107	110	100		
8. 1,784				
9. 7,842,868				

To add more than two numbers, use your addition facts. Study the example below, and then solve the addition problems using more than one step.

Find: 6 + 1 + 4 + 6

Find the sum of each set of two numbers. Write the sum beside the problem.	Add the new numbers. Line up the digits. Add the ones. Add the tens.	Write the sum.
6 >7 1 4 >10 +6	T | O | 7 +1 | 0 1 | 7	6 >7 1 4 >10 +6 17

Find the sums.

1.
| 5 >9
4
+3 3
12 | 4
4
+5 | 3
6
+2 | 7
2
+1 | 6
3
+6 | 4
5
+8 | 8
1
+3 | 2
7
+6 | 5
3
+8 |

2.
| 4
9
+6 | 5
7
+3 | 8
1
+4 | 2
7
+5 | 1
4
+8 | 3
5
+7 | 4
3
+6 | 6
5
+8 | 7
5
+6 |

3.
| 5 >7
2
5 >12
+7
19 | 3
4
6
+5 | 5
3
4
+7 | 2
7
2
+8 | 4
3
1
+5 | 9
2
5
+1 | 7
3
4
+6 | 0
1
6
+7 | 1
9
8
+5 |

4.
| 1
0
9
+7 | 0
2
4
+0 | 3
5
7
+3 | 2
4
8
+5 | 0
3
8
+8 | 6
5
8
+4 | 8
3
5
+6 | 5
5
7
+4 | 7
4
3
+9 |

5.
| 7 >15
8
9 >14
+5
29 | 4
7
5
+9 | 5
9
4
+7 | 6
8
5
+9 | 4
6
8
+9 | 1
7
4
+5 | 5
6
3
+7 | 6
8
4
+9 | 3
0
5
+7 |

6.
| 1 >3
2
4 >11 >14
7
+0 0 0
14 | 1
6
5
3
+4 | 0
4
3
6
+5 | 0
1
1
4
+9 | 4
2
2
1
+5 | 1
1
2
5
+6 | 2
3
4
1
+7 | 1
6
0
5
+0 | 3
3
1
4
+1 |

Adding Larger Numbers

To add larger numbers, start with the digits in the ones place.

Find: 524 + 163

Add the ones.	Add the tens.	Add the hundreds.
H T O	H T O	H T O
5 2 4	5 2 4	5 2 4
+ 1 6 3	+ 1 6 3	+ 1 6 3
7	8 7	6 8 7

Add.

1.
```
  33      44      37      55      82      47      70
+ 64    + 50    + 42    + 24    + 17    + 32    + 21
----    ----    ----    ----    ----    ----    ----
  97      94      79      79      99      79      91
```

2.
```
 352     432     720     630     274     322     134
+641    +345    +250    +307    +725    +475    +855
----    ----    ----    ----    ----    ----    ----
 993     777     970     937     999     797     989
```

3.
```
4,536   3,215   4,167   3,156   7,135   1,054   2,535
+3,343  +3,584  +3,522  +5,732  +2,564  +6,325  +6,442
------  ------  ------  ------  ------  ------  ------
7,879   6,799   7,689   8,888   9,699   7,379   8,977
```

4.
```
3,516   1,681   7,006   8,341   6,328   4,224   3,270
+4,102  +4,218  +2,893  +1,628  +1,050  +3,615  +4,621
------  ------  ------  ------  ------  ------  ------
7,618   5,899   9,899   9,969   7,378   7,839   7,891
```

5.
```
1,275   2,144   2,369   4,527   1,733   4,616   1,783
+3,521  +7,050  +6,530  +4,302  +6,254  +3,273  +5,214
------  ------  ------  ------  ------  ------  ------
4,796   9,194   8,899   8,829   7,987   7,889   6,997
```

6.
```
12,350  23,244  12,768  34,582  44,610  54,507
+17,629 +15,534 +16,121 +11,413 +13,071 +13,421
------- ------- ------- ------- ------- -------
29,979  38,778  28,889  45,995  57,681  67,928
```

Regrouping in Addition

To add two or more numbers, start with the digits in the ones place. Regroup as needed.

Find: 392 + 870

Add the ones.			Add the tens. Regroup.				Add the hundreds. Regroup.		

Add the ones.

Th	H	T	O
	3	9	2
+	8	7	0
			2

Add the tens. Regroup.

Th	H	T	O
	3	9	2
+	8	7	0
		6	2

To regroup, write the 6 in the tens place. Write the 1 in the hundreds column.

Add the hundreds. Regroup.

Th	H	T	O
	3	9	2
+	8	7	0
1,	2	6	2

To regroup, write the 2 in the hundreds place, write the 1 in the thousands place.

Add.

1.
```
   25      38      16      29      54      36
  +56     +43     +47     +54     +37     +45
   81      81      63      83      91      81
```

2.
```
  337     212     327     235     335     439
 +427    +359    +555    +719    +528    +432
  764     571     882     954     863     871
```

3.
```
  234     234     217     249     235     225
 +439    +358    +374    +515    +518    +226
  673     592     591     763     753     451
```

4.
```
  669     274     355     549     299     536
 +241    +347    +455    +272    +315    +274
  910     621     810     821     614     810
```

5.
```
   394    7,265   6,334    7,006   15,079
 +4,860  +1,828  +  878   +8,009  + 7,951
  5,254   9,093   7,212   15,015   23,030
```

6.
```
  234     157    2,395    6,394      236
  210      60    2,708    2,182      184
 +357    +164   +  431   +2,014   +7,706
  801     381    5,534   10,590    8,126
```

Line up the digits. Then add.

7. 386 + 424 = _____   35 + 455 = _____   1,393 + 1,870 = _____

```
   386
  +424
```

Add.

1.
```
   6        5        3        7        8        2        4        1        9
 + 5      + 0      + 8      + 2      + 0      + 6      + 4      + 3      + 8
```

2.
```
   8        7        6        7        0        9        6        9        8
   5        8        9        8        8        7        9        7        5
   9        9        7        9        0        8        7        5        9
 + 7      + 4      + 6      + 3      + 9      + 5      + 4      + 6      + 7
```

3.
```
   5 4      6 2      4 5      7 1      8 5      9 4      3 9      7 5
 + 2 3    + 1 7    + 3 2    + 2 8    + 5 7    + 4 8    + 5 2    + 9 8
```

4.
```
   3 2 4        5 1 4        6 7 3        7 9 4        6 4 5        2 6 4
 + 8 6 5      + 2 7 4      + 3 2 6      + 5 2 9      + 7 2 7      + 3 3 7
```

5.
```
   6,0 3 5      1,7 4 8      2,5 0 6      5,5 1 4      7,3 8 2      2 1,0 9 0
 + 3,7 5 2    + 9,2 2 0    + 6,1 8 2    +   4 8 2    + 5,1 0 4    + 3 1,8 0 7
```

6.
```
                                            3 9 9        4,2 0 6      1 8,0 6 0
   5,3 0 7      7,2 0 9      4,0 9 6      3,6 9 0        5,7 0 9      6 2,5 4 9
 + 6,0 9 9    + 4,0 9 5    + 8,4 0 5    + 7,0 5 9      + 6,9 9 0    +   1,3 7 5
```

Line up the digits. Then add.

7. 730 + 422 + 36 = _____   4 + 23 + 609 = _____

Solve.

8. Mihn bought 3 items at a clothing store. The items cost $26, $14, and $31. What was his total bill before sales tax?

9. Joel earns $19,700 a year. His wife earns $21,780 a year. How much do they earn altogether?

Answer _____   Answer _____

Basic Subtraction Facts

Subtraction is the opposite of addition. You subtract two numbers to find a difference. Study and master these subtraction facts.

Remember, any number minus zero is that number.

$$12 \leftarrow \text{minuend}$$
$$-\ \ 6 \leftarrow \text{subtrahend}$$
$$6 \leftarrow \text{difference}$$

Subtract.

1. $\begin{array}{r}11\\-\ 7\\\hline 4\end{array}$ $\begin{array}{r}9\\-3\\\hline\end{array}$ $\begin{array}{r}6\\-0\\\hline\end{array}$ $\begin{array}{r}16\\-\ 8\\\hline\end{array}$ $\begin{array}{r}10\\-\ 9\\\hline\end{array}$ $\begin{array}{r}13\\-\ 7\\\hline\end{array}$ $\begin{array}{r}14\\-\ 8\\\hline\end{array}$ $\begin{array}{r}15\\-\ 7\\\hline\end{array}$

2. $\begin{array}{r}14\\-\ 9\\\hline\end{array}$ $\begin{array}{r}3\\-3\\\hline\end{array}$ $\begin{array}{r}17\\-\ 8\\\hline\end{array}$ $\begin{array}{r}14\\-\ 6\\\hline\end{array}$ $\begin{array}{r}9\\-5\\\hline\end{array}$ $\begin{array}{r}11\\-\ 8\\\hline\end{array}$ $\begin{array}{r}16\\-\ 9\\\hline\end{array}$ $\begin{array}{r}9\\-6\\\hline\end{array}$

3. $\begin{array}{r}7\\-7\\\hline\end{array}$ $\begin{array}{r}12\\-\ 5\\\hline\end{array}$ $\begin{array}{r}10\\-\ 4\\\hline\end{array}$ $\begin{array}{r}12\\-\ 3\\\hline\end{array}$ $\begin{array}{r}7\\-2\\\hline\end{array}$ $\begin{array}{r}6\\-3\\\hline\end{array}$ $\begin{array}{r}12\\-\ 4\\\hline\end{array}$ $\begin{array}{r}8\\-7\\\hline\end{array}$

4. $\begin{array}{r}15\\-\ 9\\\hline\end{array}$ $\begin{array}{r}8\\-1\\\hline\end{array}$ $\begin{array}{r}18\\-\ 9\\\hline\end{array}$ $\begin{array}{r}5\\-4\\\hline\end{array}$ $\begin{array}{r}5\\-0\\\hline\end{array}$ $\begin{array}{r}12\\-\ 6\\\hline\end{array}$ $\begin{array}{r}17\\-\ 9\\\hline\end{array}$ $\begin{array}{r}13\\-13\\\hline\end{array}$

5. $\begin{array}{r}17\\-\ 7\\\hline\end{array}$ $\begin{array}{r}13\\-\ 5\\\hline\end{array}$ $\begin{array}{r}12\\-\ 7\\\hline\end{array}$ $\begin{array}{r}16\\-\ 7\\\hline\end{array}$ $\begin{array}{r}10\\-\ 6\\\hline\end{array}$ $\begin{array}{r}11\\-\ 9\\\hline\end{array}$ $\begin{array}{r}15\\-\ 8\\\hline\end{array}$ $\begin{array}{r}14\\-\ 7\\\hline\end{array}$

6. $\begin{array}{r}15\\-\ 8\\\hline\end{array}$ $\begin{array}{r}8\\-4\\\hline\end{array}$ $\begin{array}{r}6\\-1\\\hline\end{array}$ $\begin{array}{r}7\\-0\\\hline\end{array}$ $\begin{array}{r}11\\-\ 9\\\hline\end{array}$ $\begin{array}{r}9\\-0\\\hline\end{array}$ $\begin{array}{r}13\\-\ 5\\\hline\end{array}$ $\begin{array}{r}10\\-\ 7\\\hline\end{array}$

7. $\begin{array}{r}13\\-\ 4\\\hline\end{array}$ $\begin{array}{r}15\\-\ 9\\\hline\end{array}$ $\begin{array}{r}8\\-8\\\hline\end{array}$ $\begin{array}{r}11\\-\ 6\\\hline\end{array}$ $\begin{array}{r}3\\-0\\\hline\end{array}$ $\begin{array}{r}7\\-1\\\hline\end{array}$ $\begin{array}{r}10\\-\ 5\\\hline\end{array}$ $\begin{array}{r}4\\-4\\\hline\end{array}$

8. $\begin{array}{r}14\\-\ 7\\\hline\end{array}$ $\begin{array}{r}8\\-2\\\hline\end{array}$ $\begin{array}{r}16\\-\ 7\\\hline\end{array}$ $\begin{array}{r}8\\-5\\\hline\end{array}$ $\begin{array}{r}10\\-\ 2\\\hline\end{array}$ $\begin{array}{r}4\\-3\\\hline\end{array}$ $\begin{array}{r}9\\-2\\\hline\end{array}$ $\begin{array}{r}13\\-\ 6\\\hline\end{array}$

9. $\begin{array}{r}4\\-0\\\hline\end{array}$ $\begin{array}{r}18\\-\ 9\\\hline\end{array}$ $\begin{array}{r}17\\-\ 8\\\hline\end{array}$ $\begin{array}{r}12\\-\ 4\\\hline\end{array}$ $\begin{array}{r}8\\-3\\\hline\end{array}$ $\begin{array}{r}6\\-6\\\hline\end{array}$ $\begin{array}{r}10\\-\ 0\\\hline\end{array}$ $\begin{array}{r}14\\-\ 6\\\hline\end{array}$

Subtraction

To subtract, start with the digits in the ones place.

Find: 587 − 234

Subtract the ones.	Subtract the tens.	Subtract the hundreds.	
H|T|O	H|T|O	H|T|O	H|T|O
5 8 **7**	5 **8** 7	**5** 8 7	*3 5 3*
− 2 3 **4**	− 2 **3** 4	− **2** 3 4	*+ 2 3 4*
3	**5** 3	**3 5 3**	*5 8 7*

Find the differences. Check your answers.

1.

$$\begin{array}{r} 7 \\ -5 \\ \hline 2 \end{array} \quad \begin{array}{r} 9 \\ -2 \\ \hline \end{array} \quad \begin{array}{r} 6 \\ -2 \\ \hline \end{array} \quad \begin{array}{r} 8 \\ -3 \\ \hline \end{array} \quad \begin{array}{r} 7 \\ -4 \\ \hline \end{array} \quad \begin{array}{r} 9 \\ -5 \\ \hline \end{array} \quad \begin{array}{r} 8 \\ -5 \\ \hline \end{array} \quad \begin{array}{r} 7 \\ -3 \\ \hline \end{array} \quad \begin{array}{r} 9 \\ -4 \\ \hline \end{array}$$

2.

$$\begin{array}{r} 16 \\ -\ 9 \\ \hline 7 \end{array} \quad \begin{array}{r} 17 \\ -\ 8 \\ \hline \end{array} \quad \begin{array}{r} 14 \\ -\ 9 \\ \hline \end{array} \quad \begin{array}{r} 15 \\ -\ 6 \\ \hline \end{array} \quad \begin{array}{r} 16 \\ -\ 7 \\ \hline \end{array} \quad \begin{array}{r} 17 \\ -\ 9 \\ \hline \end{array} \quad \begin{array}{r} 13 \\ -\ 8 \\ \hline \end{array} \quad \begin{array}{r} 14 \\ -\ 5 \\ \hline \end{array}$$

3.

$$\begin{array}{r} 58 \\ -33 \\ \hline 25 \end{array} \quad \begin{array}{r} 87 \\ -53 \\ \hline \end{array} \quad \begin{array}{r} 79 \\ -45 \\ \hline \end{array} \quad \begin{array}{r} 67 \\ -43 \\ \hline \end{array} \quad \begin{array}{r} 98 \\ -75 \\ \hline \end{array} \quad \begin{array}{r} 87 \\ -34 \\ \hline \end{array} \quad \begin{array}{r} 45 \\ -21 \\ \hline \end{array} \quad \begin{array}{r} 76 \\ -35 \\ \hline \end{array}$$

4.

$$\begin{array}{r} 59 \\ -30 \\ \hline 29 \end{array} \quad \begin{array}{r} 94 \\ -50 \\ \hline \end{array} \quad \begin{array}{r} 52 \\ -20 \\ \hline \end{array} \quad \begin{array}{r} 76 \\ -40 \\ \hline \end{array} \quad \begin{array}{r} 98 \\ -40 \\ \hline \end{array} \quad \begin{array}{r} 65 \\ -20 \\ \hline \end{array} \quad \begin{array}{r} 67 \\ -20 \\ \hline \end{array} \quad \begin{array}{r} 23 \\ -10 \\ \hline \end{array}$$

5.

$$\begin{array}{r} 696 \\ -251 \\ \hline 445 \end{array} \quad \begin{array}{r} 995 \\ -452 \\ \hline \end{array} \quad \begin{array}{r} 877 \\ -342 \\ \hline \end{array} \quad \begin{array}{r} 788 \\ -435 \\ \hline \end{array} \quad \begin{array}{r} 987 \\ -253 \\ \hline \end{array} \quad \begin{array}{r} 579 \\ -234 \\ \hline \end{array}$$

6.

$$\begin{array}{r} 1{,}576 \\ -\ \ 834 \\ \hline 742 \end{array} \quad \begin{array}{r} 1{,}497 \\ -\ \ 943 \\ \hline \end{array} \quad \begin{array}{r} 1{,}769 \\ -\ \ 825 \\ \hline \end{array} \quad \begin{array}{r} 1{,}398 \\ -\ \ 745 \\ \hline \end{array} \quad \begin{array}{r} 1{,}687 \\ -\ \ 934 \\ \hline \end{array} \quad \begin{array}{r} 1{,}275 \\ -\ \ 542 \\ \hline \end{array}$$

7.

$$\begin{array}{r} 807 \\ -304 \\ \hline 503 \end{array} \quad \begin{array}{r} 1{,}295 \\ -\ \ 801 \\ \hline \end{array} \quad \begin{array}{r} 1{,}769 \\ -\ \ 804 \\ \hline \end{array} \quad \begin{array}{r} 958 \\ -403 \\ \hline \end{array} \quad \begin{array}{r} 769 \\ -405 \\ \hline \end{array} \quad \begin{array}{r} 896 \\ -450 \\ \hline \end{array}$$

Regrouping in Subtraction

To subtract, start with the digits in the ones place. When one or more digits in the subtrahend are larger than those in the minuend, it is necessary to *regroup* the digit in the minuend before carrying out the subtraction.

Find: 630 − 297

Regroup to subtract the ones.		Regroup again to subtract the tens.		Subtract the hundreds.	Check:

To regroup, cross out the 3 and write a 2 above it. Cross out the zero and write a 10 above it. Subtract.

```
   2 10
 6 3̷ 0̷
-2 9  7
      3
```

To regroup, cross out the 6 and write a 5 above it. Cross out the 2 and write a 12 above it. Subtract.

```
     12
  5  2̷ 10
 6̷ 3̷ 0̷
-2 9  7
    3 3
```

```
      12
  5  2̷ 10
 6̷ 3̷ 0̷
-2 9  7
 3 3  3
```

```
 1  1
 3 3 3
+2 9 7
 6 3 0
```

Subtract. Check your answers.

1.
```
  6 14
  7 4̷
 -2 5
  4 9
```
```
 8 5
-3 6
```
```
 7 5
-4 9
```
```
 8 6
-3 9
```
```
 9 6
-4 7
```
```
 5 7
-3 8
```
```
 5 3
-2 9
```
```
 7 2
-3 4
```

2.
```
    13
  5 3̷ 10
  6 4̷ 0̷
 -2 5 8
  3 8 2
```
```
 7 5 2
-2 6 9
```
```
 8 4 0
-5 4 4
```
```
 6 2 5
-3 4 9
```
```
 5 1 3
-2 0 4
```
```
 6 3 5
-2 4 7
```
```
 7 5 4
-2 7 5
```

3.
```
  16 16
  6̷ 6̷ 16
 1,7̷ 7̷ 6̷
 - 7 9 8
   9 7 8
```
```
 1,3 9 6
 -  4 7 9
```
```
 1,6 5 5
 -  7 0 7
```
```
 1,5 6 4
 -  8 4 9
```
```
 1,2 9 3
 -  4 5 4
```
```
 1,8 6 0
 -  9 4 5
```
```
 1,4 7 2
 -  9 5 8
```

4.
```
 1,2 3 2
 -  5 8 4
```
```
 1,6 7 8
 -  9 4 5
```
```
 1,4 9 8
 -  9 5 3
```
```
 1,5 8 0
 -  8 5 5
```
```
 1,6 7 7
 -  7 3 4
```
```
 1,7 7 6
 -  9 0 2
```
```
 1,2 6 5
 -  4 5 2
```

5.
```
      14
   7 4̷ 10
 8,5 0̷ 3̷
-7,6 5 2
   8 5 1
```
```
 1,0 3 2
-1,0 2 7
```
```
 3,5 0 4
-2,0 7 2
```
```
 5,7 8 3
-4,5 0 7
```
```
 3,9 2 0
-2,7 6 0
```
```
 4,7 8 0
-3,9 9 9
```
```
 7,0 6 5
-6,1 3 0
```

Subtracting Across Zeros

When subtracting from zero, you may have to regroup twice before you subtract.

Find: 503 − 234

Not enough ones. There are no tens. Regroup hundreds and tens.	Subtract the ones.	Subtract the tens.	Subtract the hundreds.	Check:

	H	T	O			H	T	O			H	T	O			H	T	O			H	T	O
		9					9					9					9				1	1	
	4	10	13			4	10	13			4	10	13			4	10	13			2	6	9
	5	0	3			5	0	3			5	0	3			5	0	3		+	2	3	4
−	2	3	4		−	2	3	4		−	2	3	4		−	2	3	4			5	0	3
								9				6	9			2	6	9					

Subtract. Check your answers.

1.
```
      9
    2 10 18
     3 0 8        6 0 4        4 0 7        3 0 3        2 0 7        6 0 3        5 0 6
   − 2 9 9      − 4 5 9      − 2 3 9      −   6 4      − 1 1 8      −   4 7      − 1 8 9
         9
```

2.
```
     5 0 7        8 0 4        4 0 3        3 0 2        9 0 1        3 0 4        8 0 2
   − 1 0 9      − 3 2 6      −   1 6      − 2 0 5      − 3 5 7      − 1 0 6      −   5 6
```

3.
```
     6 0 0        4 0 0        7 0 0        8 0 0        9 0 0        5 0 0        3 0 0
   − 1 2 4      − 2 0 8      − 5 7 0      −   7 5      − 7 0 0      − 3 0 2      −   9 6
```

4.
```
     1 0 0        3 0 0        5 0 0        2 0 0      3,6 0 0      2,7 0 0      1,8 0 0
   −   5 7      − 1 0 8      − 2 2 2      − 1 7 0      − 1,1 3 5    − 1,4 0 3    −   2 6 4
```

Line up the digits. Then subtract.

5. 301 − 25 = _____   401 − 230 = _____   300 − 142 = _____

```
   3 0 1
 −   2 5
```

Estimating Sums and Differences

To estimate, first round each number to the same place. Then add or subtract the rounded numbers. Accurate estimating will help you decide if an answer is reasonable.

Estimate: 856 + 431

Round each number to the nearest hundred. Add.
$856 \rightarrow \quad 900$ $+431 \rightarrow +400$ $\overline{\quad\quad\quad 1,300}$

Estimate: 1,583 − 632

Round each number to the nearest hundred. Subtract.
$1,583 \rightarrow \quad 1,600$ $\quad -632 \rightarrow - \quad 600$ $\overline{\quad\quad\quad 1,000}$

Estimate the sum. Round each number to the nearest hundred. Add.

1. $327 \rightarrow \quad 300$ $253 \rightarrow$ $384 \rightarrow$ $265 \rightarrow$
    $+492 \rightarrow +500$ $+485 \rightarrow$ $+234 \rightarrow$ $+341 \rightarrow$
    $\quad\quad\quad\quad 800$

2. $728 \rightarrow$ $452 \rightarrow$ $167 \rightarrow$ $272 \rightarrow$
    $+371 \rightarrow$ $+285 \rightarrow$ $+462 \rightarrow$ $+594 \rightarrow$

3. $1,543 \rightarrow$ $2,754 \rightarrow$ $3,486 \rightarrow$ $4,535 \rightarrow$
    $+ \quad 488 \rightarrow$ $+ \quad 376 \rightarrow$ $+ \quad 540 \rightarrow$ $+ \quad 480 \rightarrow$

Estimate the difference. Round each number to the nearest hundred. Subtract.

4. $482 \rightarrow \quad 500$ $357 \rightarrow$ $568 \rightarrow$ $845 \rightarrow$
    $-246 \rightarrow -200$ $-129 \rightarrow$ $-374 \rightarrow$ $-659 \rightarrow$
    $\quad\quad\quad\quad 300$

5. $682 \rightarrow$ $453 \rightarrow$ $376 \rightarrow$ $928 \rightarrow$
    $-249 \rightarrow$ $-326 \rightarrow$ $-165 \rightarrow$ $-572 \rightarrow$

6. $3,546 \rightarrow$ $2,765 \rightarrow$ $1,430 \rightarrow$ $6,828 \rightarrow$
    $- \quad 366 \rightarrow$ $- \quad 249 \rightarrow$ $- \quad 357 \rightarrow$ $- \quad 680 \rightarrow$

Checking Up

Subtract. Check your answers.

1.
$$\begin{array}{r} 8 \\ -5 \\ \hline \end{array} \qquad \begin{array}{r} 6 \\ -3 \\ \hline \end{array} \qquad \begin{array}{r} 10 \\ -9 \\ \hline \end{array} \qquad \begin{array}{r} 12 \\ -4 \\ \hline \end{array} \qquad \begin{array}{r} 16 \\ -7 \\ \hline \end{array} \qquad \begin{array}{r} 11 \\ -6 \\ \hline \end{array} \qquad \begin{array}{r} 15 \\ -8 \\ \hline \end{array} \qquad \begin{array}{r} 17 \\ -9 \\ \hline \end{array}$$

2.
$$\begin{array}{r} 39 \\ -15 \\ \hline \end{array} \qquad \begin{array}{r} 58 \\ -20 \\ \hline \end{array} \qquad \begin{array}{r} 29 \\ -19 \\ \hline \end{array} \qquad \begin{array}{r} 58 \\ -24 \\ \hline \end{array} \qquad \begin{array}{r} 73 \\ -34 \\ \hline \end{array} \qquad \begin{array}{r} 30 \\ -27 \\ \hline \end{array} \qquad \begin{array}{r} 67 \\ -18 \\ \hline \end{array} \qquad \begin{array}{r} 71 \\ -32 \\ \hline \end{array}$$

3.
$$\begin{array}{r} 896 \\ -385 \\ \hline \end{array} \qquad \begin{array}{r} 698 \\ -487 \\ \hline \end{array} \qquad \begin{array}{r} 399 \\ -108 \\ \hline \end{array} \qquad \begin{array}{r} 467 \\ -214 \\ \hline \end{array} \qquad \begin{array}{r} 704 \\ -325 \\ \hline \end{array} \qquad \begin{array}{r} 406 \\ -129 \\ \hline \end{array} \qquad \begin{array}{r} 400 \\ -208 \\ \hline \end{array}$$

4.
$$\begin{array}{r} 3{,}714 \\ -2{,}312 \\ \hline \end{array} \qquad \begin{array}{r} 7{,}413 \\ -3{,}201 \\ \hline \end{array} \qquad \begin{array}{r} 5{,}684 \\ -2{,}342 \\ \hline \end{array} \qquad \begin{array}{r} 4{,}817 \\ -2{,}805 \\ \hline \end{array} \qquad \begin{array}{r} 6{,}435 \\ -2{,}323 \\ \hline \end{array} \qquad \begin{array}{r} 1{,}942 \\ -731 \\ \hline \end{array}$$

5.
$$\begin{array}{r} 2{,}900 \\ -346 \\ \hline \end{array} \qquad \begin{array}{r} 6{,}500 \\ -1{,}234 \\ \hline \end{array} \qquad \begin{array}{r} 1{,}604 \\ -390 \\ \hline \end{array} \qquad \begin{array}{r} 4{,}290 \\ -3{,}306 \\ \hline \end{array} \qquad \begin{array}{r} 3{,}900 \\ -2{,}450 \\ \hline \end{array} \qquad \begin{array}{r} 1{,}200 \\ -999 \\ \hline \end{array}$$

6.
$$\begin{array}{r} 73{,}050 \\ -27{,}455 \\ \hline \end{array} \qquad \begin{array}{r} 46{,}940 \\ -24{,}946 \\ \hline \end{array} \qquad \begin{array}{r} 19{,}053 \\ -8{,}954 \\ \hline \end{array} \qquad \begin{array}{r} 23{,}006 \\ -4{,}999 \\ \hline \end{array} \qquad \begin{array}{r} 36{,}174 \\ -16{,}925 \\ \hline \end{array} \qquad \begin{array}{r} 86{,}502 \\ -26{,}590 \\ \hline \end{array}$$

Line up the digits. Then subtract.

7. $6{,}030 - 429 =$ _____ $399 - 62 =$ _____ $700 - 30 =$ _____

Solve.

8. A magazine has 34,783 subscribers. Last year the magazine had 26,876 subscribers. By how much did the number of subscribers increase?

9. The first printing of a book sold 45,421 copies. The second issue of the book sold 56,783 copies. How many more second-issue books were sold?

Answer _____

Answer _____

Problem-Solving Strategy: Choose an Operation

Sometimes a problem does not tell you whether to add or subtract, multiply or divide. To solve such a problem, you must read the problem carefully. Then, decide what the problem is asking you to do. Next, choose an operation and solve the problem. Watch for clue words such as: **in all, total, how many** . . . ?, or **how many more** . . . ?

STEPS

1. Read the problem.

> The three heaviest players on a football team weigh 290 pounds, 278 pounds, and 304 pounds. What is the total weight of these three players?

2. Decide what the problem is asking.

> In this problem, the question "What is the **total** . . . ?" is asking you to find a sum.

3. Choose the operation.

> To solve, you must add.

4. Solve the problem.

> $290 + 278 + 304 = 872$
>
> The **total** weight of the three players is 872 pounds.

Choose the correct operation and solve each problem.

1. The distance from Chicago, Illinois to Butte, Montana is 1,522 miles. Seattle, Washington is 567 miles beyond Butte. **How far** is it from Chicago to Seattle?

Operation _____

Answer _____

2. Jerome counted 845 old books and 519 new books for the book sale. **How many more** old books were at the sale than new books?

Operation _____

Answer _____

3. Fletcher cashed a check for $292. He deposited $105 of that money in his account. **How much** of the $292 did Fletcher keep?

Operation _____

Answer _____

4. Sue scored a 78, an 81, a 76, and a 79 in the golf tournament. What was her **total** score?

Operation _____

Answer _____

Applying Your Skills

Choose an operation and solve.
Watch for clue words.

1. Marie budgeted the following for her monthly bills: $450 for rent and $100 for electricity. **How much** did she budget altogether for these bills? _____

2. On one Saturday, 73,927 people attended a basketball game. The next week 68,412 people saw the game. **How many** people saw the two games? _____

3. The area of North Carolina is 48,843 square miles. South Carolina's area is 30,207 square miles. Which is larger? _____
   How much larger? _____

4. The Cascade Tunnel in Washington is 41,152 feet long. The Moffat Tunnel in Colorado is 32,798 feet long. **How much longer** is the Cascade Tunnel than the Moffat Tunnel? _____

5. In one year 28,762 new books were published. The next year 30,387 books were published. **How many** books were published during the two years? _____

6. Apollo 8 flew 550,000 miles on its trip around the moon. Apollo 9 flew 3,700,000 miles on its trip. **How many** miles were flown by the two spacecraft? _____

7. The population of Arkansas is 2,296,419. _____
   Arizona has 2,718,425 people. Which has more people? **How many more?** _____

8. Juan and Ho combined their baseball card collection. Juan has 1,094 cards. Together they have 2,106 cards. **How many** cards does Ho have? _____

9. The attendance at Marshall High School's football game one week was 5,479. The next week 4,388 students attended the game. Altogether, **how many** people attended these two games? _____

10. Mount Whitney is 14,494 feet high. Borah Peak is 12,662 feet high. **How much higher** is Mount Whitney? _____

Basic Multiplication Facts

Multiplication is a short way to do addition.

4×5 means 4 *times* 5 or $5 + 5 + 5 + 5 = 20$.
5×4 means 5 *times* 4 or $4 + 4 + 4 + 4 + 4 = 20$.

The order in which numbers are multiplied does not change the answer. The answer is called the **product**.

4×5 can be written

$$\begin{array}{r} 5 \\ \times 4 \\ \hline 20 \end{array}$$

5×4 can be written

$$\begin{array}{rl} 4 & \leftarrow \text{multiplicand} \\ \times 5 & \leftarrow \text{multiplier} \\ \hline 20 & \leftarrow \text{product} \end{array}$$

Remember,
- zero times any number is zero.
- one times any number is that number.

The multiplication table below shows each number from 0 through 9 multiplied by each number from 0 through 9. Study the multiplication facts.

X	0	1	2	3	4	5	6	7	8	9
0	0	0	0	0	0	0	0	0	0	0
1	0	1	2	3	4	5	6	7	8	9
2	0	2	4	6	8	10	12	14	16	18
3	0	3	6	9	12	15	18	21	24	27
4	0	4	8	12	16	20	24	28	32	36
5	0	5	10	15	20	25	30	35	40	45
6	0	6	12	18	24	30	36	42	48	54
7	0	7	14	21	28	35	42	49	56	63
8	0	8	16	24	32	40	48	56	64	72
9	0	9	18	27	36	45	54	63	72	81

Multiply. Use the multiplication table.

1. $6 \times 4 = \underline{\quad 24 \quad}$ $7 \times 2 = \underline{\qquad}$ $5 \times 5 = \underline{\qquad}$ $5 \times 3 = \underline{\qquad}$

2. $3 \times 2 = \underline{\qquad}$ $2 \times 8 = \underline{\qquad}$ $8 \times 3 = \underline{\qquad}$ $0 \times 3 = \underline{\qquad}$

3. $9 \times 5 = \underline{\qquad}$ $4 \times 7 = \underline{\qquad}$ $6 \times 8 = \underline{\qquad}$ $7 \times 9 = \underline{\qquad}$

4. $8 \times 5 = \underline{\qquad}$ $9 \times 6 = \underline{\qquad}$ $4 \times 9 = \underline{\qquad}$ $2 \times 6 = \underline{\qquad}$

5. $6 \times 7 = \underline{\qquad}$ $7 \times 8 = \underline{\qquad}$ $5 \times 6 = \underline{\qquad}$ $3 \times 3 = \underline{\qquad}$

Basic Multiplication Facts

Study and master these multiplication facts.

Multiply.

1. $\begin{array}{r} 5 \\ \times\,2 \\ \hline \end{array}$ *10* $\quad$ $\begin{array}{r} 7 \\ \times\,2 \\ \hline \end{array}$ $\quad$ $\begin{array}{r} 9 \\ \times\,0 \\ \hline \end{array}$ $\quad$ $\begin{array}{r} 1 \\ \times\,3 \\ \hline \end{array}$ $\quad$ $\begin{array}{r} 5 \\ \times\,4 \\ \hline \end{array}$ $\quad$ $\begin{array}{r} 4 \\ \times\,3 \\ \hline \end{array}$ $\quad$ $\begin{array}{r} 6 \\ \times\,0 \\ \hline \end{array}$ $\quad$ $\begin{array}{r} 9 \\ \times\,4 \\ \hline \end{array}$

2. $\begin{array}{r} 8 \\ \times\,5 \\ \hline \end{array}$ $\quad$ $\begin{array}{r} 3 \\ \times\,5 \\ \hline \end{array}$ $\quad$ $\begin{array}{r} 7 \\ \times\,0 \\ \hline \end{array}$ $\quad$ $\begin{array}{r} 6 \\ \times\,4 \\ \hline \end{array}$ $\quad$ $\begin{array}{r} 2 \\ \times\,8 \\ \hline \end{array}$ $\quad$ $\begin{array}{r} 5 \\ \times\,0 \\ \hline \end{array}$ $\quad$ $\begin{array}{r} 1 \\ \times\,8 \\ \hline \end{array}$ $\quad$ $\begin{array}{r} 9 \\ \times\,5 \\ \hline \end{array}$

3. $\begin{array}{r} 3 \\ \times\,4 \\ \hline \end{array}$ $\quad$ $\begin{array}{r} 0 \\ \times\,8 \\ \hline \end{array}$ $\quad$ $\begin{array}{r} 6 \\ \times\,3 \\ \hline \end{array}$ $\quad$ $\begin{array}{r} 9 \\ \times\,1 \\ \hline \end{array}$ $\quad$ $\begin{array}{r} 9 \\ \times\,2 \\ \hline \end{array}$ $\quad$ $\begin{array}{r} 5 \\ \times\,6 \\ \hline \end{array}$ $\quad$ $\begin{array}{r} 4 \\ \times\,6 \\ \hline \end{array}$ $\quad$ $\begin{array}{r} 3 \\ \times\,8 \\ \hline \end{array}$

4. $\begin{array}{r} 1 \\ \times\,7 \\ \hline \end{array}$ $\quad$ $\begin{array}{r} 4 \\ \times\,0 \\ \hline \end{array}$ $\quad$ $\begin{array}{r} 9 \\ \times\,6 \\ \hline \end{array}$ $\quad$ $\begin{array}{r} 3 \\ \times\,3 \\ \hline \end{array}$ $\quad$ $\begin{array}{r} 5 \\ \times\,8 \\ \hline \end{array}$ $\quad$ $\begin{array}{r} 7 \\ \times\,3 \\ \hline \end{array}$ $\quad$ $\begin{array}{r} 0 \\ \times\,2 \\ \hline \end{array}$ $\quad$ $\begin{array}{r} 4 \\ \times\,8 \\ \hline \end{array}$

5. $\begin{array}{r} 0 \\ \times\,6 \\ \hline \end{array}$ $\quad$ $\begin{array}{r} 6 \\ \times\,6 \\ \hline \end{array}$ $\quad$ $\begin{array}{r} 7 \\ \times\,7 \\ \hline \end{array}$ $\quad$ $\begin{array}{r} 8 \\ \times\,8 \\ \hline \end{array}$ $\quad$ $\begin{array}{r} 1 \\ \times\,9 \\ \hline \end{array}$ $\quad$ $\begin{array}{r} 4 \\ \times\,7 \\ \hline \end{array}$ $\quad$ $\begin{array}{r} 0 \\ \times\,0 \\ \hline \end{array}$ $\quad$ $\begin{array}{r} 6 \\ \times\,7 \\ \hline \end{array}$

6. $\begin{array}{r} 9 \\ \times\,9 \\ \hline \end{array}$ $\quad$ $\begin{array}{r} 9 \\ \times\,7 \\ \hline \end{array}$ $\quad$ $\begin{array}{r} 0 \\ \times\,5 \\ \hline \end{array}$ $\quad$ $\begin{array}{r} 9 \\ \times\,3 \\ \hline \end{array}$ $\quad$ $\begin{array}{r} 8 \\ \times\,1 \\ \hline \end{array}$ $\quad$ $\begin{array}{r} 8 \\ \times\,4 \\ \hline \end{array}$ $\quad$ $\begin{array}{r} 8 \\ \times\,6 \\ \hline \end{array}$ $\quad$ $\begin{array}{r} 8 \\ \times\,0 \\ \hline \end{array}$

7. $\begin{array}{r} 1 \\ \times\,6 \\ \hline \end{array}$ $\quad$ $\begin{array}{r} 8 \\ \times\,9 \\ \hline \end{array}$ $\quad$ $\begin{array}{r} 1 \\ \times\,0 \\ \hline \end{array}$ $\quad$ $\begin{array}{r} 8 \\ \times\,3 \\ \hline \end{array}$ $\quad$ $\begin{array}{r} 7 \\ \times\,8 \\ \hline \end{array}$ $\quad$ $\begin{array}{r} 4 \\ \times\,9 \\ \hline \end{array}$ $\quad$ $\begin{array}{r} 7 \\ \times\,5 \\ \hline \end{array}$ $\quad$ $\begin{array}{r} 0 \\ \times\,9 \\ \hline \end{array}$

8. $\begin{array}{r} 0 \\ \times\,7 \\ \hline \end{array}$ $\quad$ $\begin{array}{r} 5 \\ \times\,9 \\ \hline \end{array}$ $\quad$ $\begin{array}{r} 3 \\ \times\,9 \\ \hline \end{array}$ $\quad$ $\begin{array}{r} 7 \\ \times\,1 \\ \hline \end{array}$ $\quad$ $\begin{array}{r} 4 \\ \times\,5 \\ \hline \end{array}$ $\quad$ $\begin{array}{r} 7 \\ \times\,6 \\ \hline \end{array}$ $\quad$ $\begin{array}{r} 7 \\ \times\,9 \\ \hline \end{array}$ $\quad$ $\begin{array}{r} 6 \\ \times\,9 \\ \hline \end{array}$

9. $\begin{array}{r} 9 \\ \times\,8 \\ \hline \end{array}$ $\quad$ $\begin{array}{r} 8 \\ \times\,7 \\ \hline \end{array}$ $\quad$ $\begin{array}{r} 4 \\ \times\,4 \\ \hline \end{array}$ $\quad$ $\begin{array}{r} 5 \\ \times\,7 \\ \hline \end{array}$ $\quad$ $\begin{array}{r} 2 \\ \times\,0 \\ \hline \end{array}$ $\quad$ $\begin{array}{r} 2 \\ \times\,2 \\ \hline \end{array}$ $\quad$ $\begin{array}{r} 6 \\ \times\,5 \\ \hline \end{array}$ $\quad$ $\begin{array}{r} 6 \\ \times\,8 \\ \hline \end{array}$

Multiplying Larger Numbers

To multiply, line up the digits. Multiply, starting with the digits in the ones place. Write a zero as a placeholder. Add to find the final product.

Remember,
- zero times a number is zero.
- one times a number is that number.

Find: 34 × 2

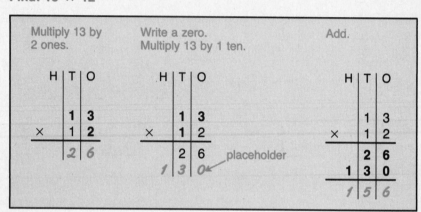

Multiply the 4 by 2 ones.	Multiply the 30 by 2 ones.
T|O 3|4 × |2 — |8	T|O 3|4 × |2 — 6|8

Find: 13 × 12

Multiply 13 by 2 ones.	Write a zero. Multiply 13 by 1 ten.	Add.
H|T|O |1|3 × |1|2 — |2|6	H|T|O |1|3 × |1|2 — |2|6 1|3|0 ← placeholder	H|T|O |1|3 × |1|2 — |2|6 1|3|0 — 1|5|6

Multiply.

1.
```
  12      22      34      20      11      24      30      21
×  4    ×  3    ×  2    ×  3    ×  7    ×  2    ×  3    ×  3
————
  48
```

2.
```
  21      31      10      23      14      33      20      11
×  4    ×  3    ×  8    ×  3    ×  2    ×  3    ×  4    ×  5
————
  84
```

3.
```
  12      12      23      33      42      61      24      44
×12     ×13     ×20     ×33     ×21     ×10     ×22     ×22
————
  24
 120
————
 144
```

4.
```
  234      123      302      434      413      210
×  12    ×  32    ×  23    ×  22    ×  23    ×  43
————
  468
 2340
——————
 2,808
```

5.
```
  2,332      1,202      3,113      4,231      1,110
×   123    ×   321    ×   230    ×   212    ×   213
————————
    6996
   46640
  233200
—————————
  286,836
```

22

Regrouping in Multiplication

To multiply by one-digit numbers, use your basic multiplication facts. It is sometimes necessary to regroup when multiplying.

Find: 4 × 37

Multiply 7 by 4 ones. Regroup ones.

H	T	O
	2	
3	7	
×		4
		8

7 × 4 = 28
To regroup, put the 8 in the ones place. Put the 2 in the tens column.

Multiply 30 by 4 ones. Regroup tens.

H	T	O
	2	
3	7	
×		4
1	4	8

3 × 4 = 12
12 + 2 = 14
Put the 4 in the tens place. Put the 1 in the hundreds column.

Multiply.

1.
```
    94        50        32        24        72        60
  ×  2      ×  5      ×  4      ×  3      ×  2      ×  2
  ─────
   188
```

2.
```
    40        93        60        64        71        43
  ×  8      ×  2      ×  6      ×  2      ×  2      ×  3
```

3.
```
     1
   653       531       600       790       540       432
  ×   3     ×   6     ×   4     ×   3     ×   4     ×   4
  ─────
  1,959
```

4.
```
   642       881       560       670       472       362
  ×   4     ×   3     ×   4     ×   4     ×   4     ×   4
```

5.
```
   2 2
   544       904       342       429       296       433
  ×   6     ×   4     ×   7     ×   9     ×   8     ×   9
  ─────
  3,264
```

6.
```
   607       458       760       253       404       850
  ×   4     ×   4     ×   5     ×   6     ×   5     ×   4
```

7.
```
  1,730     2,606     3,743     1,450     2,354     3,647
  ×    4    ×    5    ×    3    ×    7    ×    4    ×    6
```

More Regrouping in Multiplication

To multiply by two-digit or three-digit numbers, use your multiplication facts. Line up the digits. Multiply, starting with the digits in the ones place. Regroup if needed. Write a zero as a placeholder. Add to find the product. Remember, zero times a number is zero. One times a number is that number.

Find: 36 × 45

	Multiply by 6 ones. Regroup.	Write a zero. Multiply by 3 tens. Regroup.	Add.

Multiply by 6 ones. Regroup.

Th	H	T	O
		3	
		4	5
×		3	6
	2	7	0

Write a zero. Multiply by 3 tens. Regroup.

Th	H	T	O
		1	
		4	5
×		3	6
	2	7	0
1	3	5	0	← placeholder

Add.

Th	H	T	O
		4	5
×		3	6
	2	7	0
1	3	5	0
1,	6	2	0

Multiply.

1.
```
   43        56        65        72        53        87        45        36
 × 19      × 27      × 35      × 18      × 14      × 23      × 32      × 33
  387
  430
  817
```

2.
```
   79        86        77        93        62        43        36        29
 × 22      × 16      × 85      × 16      × 48      × 53      × 28      × 75
  158
 1580
 1,738
```

3.
```
   155       268       370       259       380       774       938
 ×  37     ×  25     ×  44     ×  22     ×  23     ×  34     ×  46
  1085
  4650
  5,735
```

4.
```
  1,618     2,494     1,527     2,632     2,764     3,440     2,925
 ×    34   ×    24   ×    59   ×    49   ×    28   ×    33   ×    75
   6472
  48540
  55,012
```

Line up the digits. Then multiply.

5. 217 × 35 = _____   905 × 24 = _____   25 × 32 = _____

```
    217
 ×   35
```

Zeros in Multiplication

Remember,
* the product of zero and any number is zero.
* the sum of zero and any number is that number.

Find: 20 × 608

Multiply by 0 ones.	Write a zero as a placeholder. Multiply by 2 tens. Regroup.	Add.

TTh	Th	H	T	O
		6	0	8
×			2	0
0	0	0	0	

TTh	Th	H	T	O
			1	
		6	0	8
×			2	0
	0	0	0	0
1	2	1	6	0

← placeholder

TTh	Th	H	T	O
		6	0	8
×			2	0
	0	0	0	0
1	2	1	6	0
1	2,	1	6	0

Multiply.

1.
```
  30        20        10        60        40        50
×  3      ×  6      ×  7      ×  4      ×  6      ×  8
────                                                  
  90
```

2.
```
  709       304       580       102       603       906
×  82     ×  49     ×  21     ×  98     ×  37     ×  53
─────                                                 
 1418 ↙
56720
─────
58,138
```

3.
```
  125       365       455       248       914       779
×  50     ×  10     ×  70     ×  20     ×  30     ×  40
─────                                                 
 000 ↙
6250
─────
6,250
```

4.
```
  507       806       390       204       109       603
×  20     ×  40     ×  60     ×  90     ×  20     ×  80
─────                                                 
  000 ↙
10140
─────
10,140
```

5.
```
  600       200       500       400       300       900
×  70     ×  30     ×  60     ×  50     ×  70     ×  30
─────                                                 
  000 ↙
42000
─────
42,000
```

Multiply.

1.
$$48 \times 7 \qquad 50 \times 7 \qquad 78 \times 9 \qquad 90 \times 9 \qquad 86 \times 8 \qquad 67 \times 8 \qquad 49 \times 8$$

2.
$$640 \times 9 \qquad 769 \times 0 \qquad 825 \times 6 \qquad 892 \times 8 \qquad 403 \times 9 \qquad 536 \times 7$$

3.
$$785 \times 7 \qquad 849 \times 1 \qquad 670 \times 9 \qquad 935 \times 5 \qquad 680 \times 7 \qquad 324 \times 8$$

4.
$$53 \times 25 \qquad 46 \times 40 \qquad 90 \times 58 \qquad 19 \times 39 \qquad 28 \times 40 \qquad 37 \times 50 \qquad 25 \times 10$$

5.
$$805 \times 69 \qquad 400 \times 10 \qquad 930 \times 30 \qquad 764 \times 46 \qquad 300 \times 50 \qquad 310 \times 20$$

6.
$$2{,}675 \times 20 \qquad 9{,}705 \times 49 \qquad 3{,}620 \times 30 \qquad 5{,}649 \times 28 \qquad 1{,}570 \times 10 \qquad 2{,}805 \times 15$$

Line up the digits. Then multiply.

7. $300 \times 90 =$ _____ $5{,}660 \times 25 =$ _____ $99 \times 20 =$ _____

Solve.

8. A boxcar can carry 1,540 bushels of wheat. A bushel of wheat weighs 60 pounds. How many pounds of wheat can a boxcar carry?

Answer _____

9. A freight train has 70 cars. Each car can hold 50,000 pounds of material. How much weight can the train hold in all?

Answer _____

Basic Division Facts

Division is the reverse, or opposite, of multiplication. For example, in problem 1, since $2 \times 8 = 16$, you will find by reversing the process that $16 \div 8 = 2$ and $16 \div 2 = 8$.

Remember, any number multiplied by zero equals zero. When zero is divided by any number (except zero), the answer is zero. Thus, $0 \div 5 = 0$. However, division by zero is not possible. In the problem $5 \div 0$, there is no number such that $0 \times ? = 5$.

Multiply. Then divide.

1. $2 \times 8 = \underline{\quad 16 \quad}$ $16 \div 8 = \underline{\quad 2 \quad}$ $16 \div 2 = \underline{\quad 8 \quad}$

2. $4 \times 8 = \underline{\quad\quad}$ $32 \div 4 = \underline{\quad\quad}$ $32 \div 8 = \underline{\quad\quad}$

3. $5 \times 7 = \underline{\quad\quad}$ $35 \div 7 = \underline{\quad\quad}$ $35 \div 5 = \underline{\quad\quad}$

4. $3 \times 9 = \underline{\quad\quad}$ $27 \div 3 = \underline{\quad\quad}$ $27 \div 9 = \underline{\quad\quad}$

5. $5 \times 8 = \underline{\quad\quad}$ $40 \div 8 = \underline{\quad\quad}$ $40 \div 5 = \underline{\quad\quad}$

6. $3 \times 6 = \underline{\quad\quad}$ $18 \div 6 = \underline{\quad\quad}$ $18 \div 3 = \underline{\quad\quad}$

7. $4 \times 9 = \underline{\quad\quad}$ $36 \div 4 = \underline{\quad\quad}$ $36 \div 9 = \underline{\quad\quad}$

8. $5 \times 9 = \underline{\quad\quad}$ $45 \div 9 = \underline{\quad\quad}$ $45 \div 5 = \underline{\quad\quad}$

9. $4 \times 7 = \underline{\quad\quad}$ $28 \div 7 = \underline{\quad\quad}$ $28 \div 4 = \underline{\quad\quad}$

10. $5 \times 4 = \underline{\quad\quad}$ $20 \div 4 = \underline{\quad\quad}$ $20 \div 5 = \underline{\quad\quad}$

11. $5 \times 0 = \underline{\quad 0 \quad}$ $0 \div 9 = \underline{\quad\quad}$ $0 \div 5 = \underline{\quad\quad}$

12. $4 \times 0 = \underline{\quad\quad}$ $0 \div 3 = \underline{\quad\quad}$ $0 \div 4 = \underline{\quad\quad}$

Divide.

13. $5\overline{)30}$ _(6)_ $3\overline{)24}$ $4\overline{)20}$ $5\overline{)25}$ $4\overline{)24}$ $5\overline{)15}$ $4\overline{)12}$

14. $5\overline{)40}$ $4\overline{)28}$ $2\overline{)14}$ $8\overline{)40}$ $5\overline{)10}$ $3\overline{)21}$ $2\overline{)16}$

15. $3\overline{)27}$ $2\overline{)18}$ $4\overline{)32}$ $5\overline{)45}$ $4\overline{)36}$ $2\overline{)12}$ $3\overline{)9}$

16. $2\overline{)10}$ $5\overline{)5}$ $4\overline{)8}$ $3\overline{)12}$ $6\overline{)30}$ $2\overline{)6}$ $3\overline{)18}$

17. $9\overline{)27}$ $8\overline{)24}$ $6\overline{)12}$ $9\overline{)18}$ $7\overline{)14}$ $7\overline{)21}$ $9\overline{)36}$

Practicing Division

In division, the number you divide by is called the **divisor.**
The number being divided is the **dividend.** The answer is
the **quotient.**

$$\overset{8 \leftarrow \text{quotient}}{\text{divisor} \rightarrow 2\overline{)16}} \leftarrow \text{dividend}$$

Divide.

1. $2\overset{2}{\overline{)4}}$ $2\overline{)6}$ $2\overline{)8}$ $2\overline{)10}$ $2\overline{)12}$ $2\overline{)14}$

2. $3\overline{)6}$ $3\overline{)12}$ $3\overline{)18}$ $3\overline{)24}$ $3\overline{)15}$ $3\overline{)21}$

3. $4\overline{)8}$ $4\overline{)20}$ $4\overline{)32}$ $4\overline{)16}$ $4\overline{)24}$ $4\overline{)36}$

4. $6\overline{)12}$ $6\overline{)24}$ $6\overline{)18}$ $6\overline{)30}$ $6\overline{)42}$ $6\overline{)36}$

5. $7\overline{)21}$ $7\overline{)42}$ $7\overline{)35}$ $7\overline{)63}$ $7\overline{)28}$ $7\overline{)49}$

6. $5\overline{)45}$ $5\overline{)15}$ $5\overline{)25}$ $5\overline{)20}$ $5\overline{)35}$ $5\overline{)30}$

7. $8\overline{)24}$ $8\overline{)48}$ $8\overline{)16}$ $8\overline{)32}$ $8\overline{)64}$ $8\overline{)40}$

8. $9\overline{)54}$ $6\overline{)48}$ $8\overline{)72}$ $9\overline{)63}$ $4\overline{)28}$ $9\overline{)45}$

9. $6\overline{)42}$ $9\overline{)81}$ $8\overline{)56}$ $5\overline{)45}$ $3\overline{)27}$ $7\overline{)42}$

10. $7\overset{80}{\overline{)560}}$ **Think:** 7 x ? = 56 $6\overline{)546}$ $9\overline{)279}$ $9\overline{)189}$ $8\overline{)488}$
    Think: 7 x ? = 0

11. $9\overline{)180}$ $7\overline{)147}$ $6\overline{)306}$ $8\overline{)408}$ $5\overline{)255}$ $4\overline{)248}$

12. $6\overline{)426}$ $7\overline{)497}$ $8\overline{)328}$ $9\overline{)549}$ $4\overline{)120}$ $5\overline{)350}$

Long Division

Sometimes more than one step is needed to solve division problems. Study the example of **long division** below. Use your division facts.

Find: 96 ÷ 4

Step 1	T	O	
	2		**Divide:** 9 ÷ 4 is about 2
4)9	6		Write 2 above the 9.

Step 2			
	2		**Multiply:** 2 × 4 = 8
4)9	6		Write 8 under the 9.
8			

Step 3			
	2		**Subtract:** 9 − 8 = 1
4)9	6		Write 1 under the 8.
8			**Bring down:** Bring down the 6.
1	*6*		Write the 6 right after the remainder 1. This gives 16.

Step 4			
	2	*4*	**Divide:** 16 ÷ 4 = 4
4)9	6		Write the 4 above the 6.
8			**Multiply:** 4 × 4 = 16. Write the 16 under the 16 remainder.
1	6		**Subtract:** 16 − 16 = 0. There is nothing left over.
1	*6*		The division is complete.
0			

Check:	T	O	
	1		Multiply the answer by the divisor.
	2	4	The product should equal the dividend.
×		4	
	9	6	

Divide. Check your answers.

1.
```
     2 3
4 ) 9 2
    8 ↓
    1 2
    1 2
      0
```
    6)9 6   5)8 0   7)9 1   8)9 6

2. 7)8 4   5)6 5   6)7 8   3)5 1   4)6 4

29

One-digit Divisors

To divide by a one-digit divisor, use your division facts.

Find: 68 ÷ 2

Divide 2 into 6.	Multiply and subtract. Bring down the 8.	Divide 2 into 8. Multiply and subtract. The remainder is zero. The division is complete.	Check: Multiply divisor and quotient.

$$
\begin{array}{c|c}
\text{T} & \text{O} \\
\hline
2\overline{)6} & 8
\end{array}
\qquad
\begin{array}{r}
3 \\
2\overline{)6\,8} \\
\underline{6}\downarrow \\
0\,\,|\,8
\end{array}
\qquad
\begin{array}{r}
3\,|\,4 \\
2\overline{)6\,|\,8} \\
\underline{6} \\
0\,|\,8 \\
\underline{8} \\
0
\end{array}
\qquad
\begin{array}{r}
3\,4 \\
\times\ \ 2 \\
\hline
6\,8
\end{array}
$$

The product should equal the dividend.

Divide. Check your answers.

1.
$$
\begin{array}{r}
1\,6 \\
6\overline{)9\,6} \\
\underline{6}\downarrow \\
3\,6 \\
\underline{3\,6} \\
0
\end{array}
$$
  $7\overline{)84}$ $4\overline{)92}$ $5\overline{)85}$ $7\overline{)91}$ $6\overline{)84}$

2. $4\overline{)76}$ $4\overline{)72}$ $3\overline{)84}$ $8\overline{)96}$ $6\overline{)72}$ $4\overline{)96}$

3.
$$
\begin{array}{r}
5\,8 \\
6\overline{)3\,4\,8} \\
\underline{3\,0}\downarrow \\
4\,8 \\
\underline{4\,8} \\
0
\end{array}
$$
  $3\overline{)192}$ $4\overline{)172}$ $5\overline{)165}$ $6\overline{)192}$ $7\overline{)224}$

4.
$$
\begin{array}{r}
1\,4\,2 \\
4\overline{)5\,6\,8} \\
\underline{4}\downarrow \\
1\,6 \\
\underline{1\,6}\downarrow \\
0\,8 \\
\underline{8} \\
0
\end{array}
$$
  $6\overline{)726}$ $8\overline{)968}$ $7\overline{)847}$ $5\overline{)755}$ $3\overline{)423}$

Division with Remainders

Sometimes, numbers do not divide into other numbers evenly. When a number is left over after dividing, the number is called a **remainder.** The remainder is written as part of the quotient. The remainder is always smaller than the divisor.

Division problems can be checked by multiplying the quotient and the divisor. Then add the remainder to the product. The answer should equal the dividend.

Find: 437 ÷ 8

Divide 8 into 43.

Multiply and subtract. Bring down the 7.

Divide 8 into 37. Multiply and subtract. The remainder is 5. You can't divide 8 into 5 evenly.

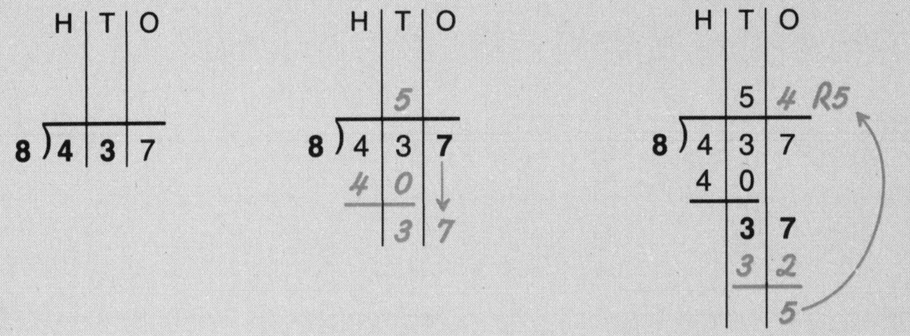

Write the remainder as part of the quotient. The division is complete.

Check: Multiply the quotient by the divisor. Add the remainder.

$$
\begin{array}{r}
54 \\
\times\ \ 8 \\
\hline
432 \\
+\ \ \ 5 \\
\hline
437
\end{array}
$$

The sum should equal the dividend.

Divide.

1. $3\overline{)70}$ → $23\ R1$

   $4\overline{)87}$ $5\overline{)56}$ $7\overline{)92}$ $6\overline{)69}$

2. $4\overline{)565}$ $6\overline{)729}$ $7\overline{)800}$ $3\overline{)434}$ $8\overline{)998}$

3. $5\overline{)148}$ → $29\ R3$

   $8\overline{)169}$ $8\overline{)253}$ $4\overline{)343}$ $3\overline{)236}$

Finding Trial Quotients

When dividing by 2-digit numbers, it is often necessary to guess the quotient. This guess is called a trial quotient. After you decide on a trial quotient, multiply and subtract. If your trial quotient is too large or too small, try again.

Find: 672 ÷ 28

Divide 28 into 67.	Multiply and subtract.	Try a smaller number. Multiply and subtract.	Finish the problem.
$28\overline{)672}$	$28\overline{)672}$ with 3 above, 84	$28\overline{)672}$ with 2 above, 56, 11	$28\overline{)672}$ with 24 above, 56, 112, 112, 0
Think: $2\overline{)6}$ is 3. So, $28\overline{)67}$ is about 3.	Since 84 > 67, 3 is too large.	Since 11 < 28, 2 is correct.	

Find: 675 ÷ 15

$15\overline{)675}$	$15\overline{)675}$ with 3 above, 45, 22	$15\overline{)675}$ with 4 above, 60, 7	$15\overline{)675}$ with 45 above, 60, 75, 75, 0
15 rounds to 20. Think: $2\overline{)6}$ is 3. So, $15\overline{)67}$ is about 3.	Since 22 > 15, 3 is too small.	Since 7 < 15, 4 is correct.	

Identify the given quotient as too large or too small. Write the correct trial quotient on the line.

1. $25\overline{)475}$ quotient 2, 50 _too large_ $15\overline{)682}$ quotient 3 _____ $18\overline{)813}$ quotient 5 _____

   _1_

2. $61\overline{)2,419}$ quotient 4 _____ $42\overline{)1,253}$ quotient 3 _____ $59\overline{)1,847}$ quotient 2 _____

3. $54\overline{)41,249}$ quotient 8 _____ $27\overline{)65,487}$ quotient 3 _____ $31\overline{)56,382}$ quotient 2 _____

32

Two-digit and Three-digit Divisors

To divide by two-digit and three-digit divisors, decide on a trial quotient. Multiply and subtract. Write the remainder in the quotient.

Find: 748 ÷ 35

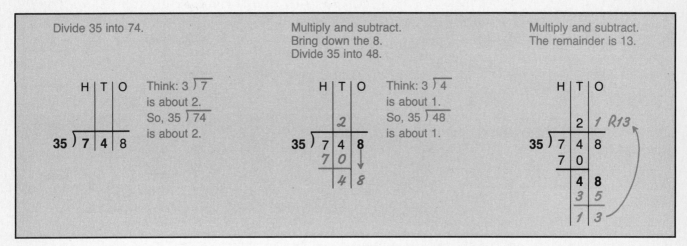

Divide 35 into 74.

Think: 3)7 is about 2. So, 35)74 is about 2.

35)7 4 8

Multiply and subtract. Bring down the 8. Divide 35 into 48.

Think: 3)4 is about 1. So, 35)48 is about 1.

Multiply and subtract. The remainder is 13.

Divide.

1. 24)5,113   → 213 R1

26)5,512

52)5,782

17)3,757

2. 72)4,128

14)7,497

32)9,984

41)5,379

3. 216)48,384   → 224

412)51,084

334)41,082

275)61,600

4. 346)46,719

213)72,633

334)54,446

122)41,724

Zeros in Quotients

To divide, decide on a trial quotient. Then multiply and subtract. Remember to divide every time you bring down a number.

Find: 66,746 ÷ 324

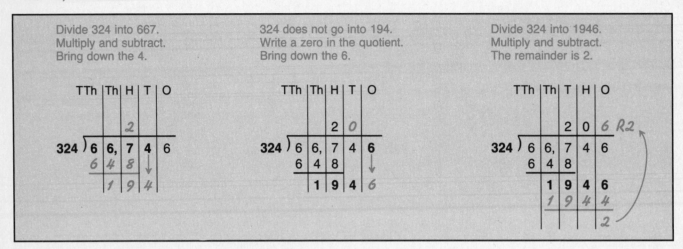

Divide 324 into 667.
Multiply and subtract.
Bring down the 4.

324 into 66,746 quotient 2, 648, 194

324 does not go into 194.
Write a zero in the quotient.
Bring down the 6.

324 into 66,746 quotient 20, 648, 1946

Divide 324 into 1946.
Multiply and subtract.
The remainder is 2.

324 into 66,746 quotient 206 R2, 648, 1946, 1944, 2

Divide.

1. 68)27,608
   406
   272
   408
   408
   0

32)98,560

15)90,015

2. 214)649,473
   3,034 R197
   642
   747
   642
   1053
   856
   197

312)939,125

413)842,945

3. 216)43,850

41)266,910

334)40,193

4. 275)556,619

418)169,290

161)484,499

Finding Averages

Averages are found by adding a set of numbers and dividing the sum by the number of items added.

Find the average of 178, 164, and 159.

Ms. Rodriguez bought groceries for four weeks and spent $91, $105, $86, and $114. Find the average amount per week she spent on groceries.

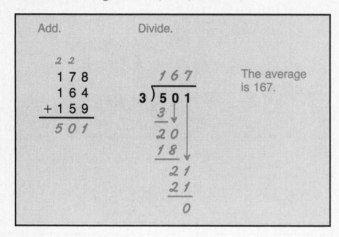

The average is 167.

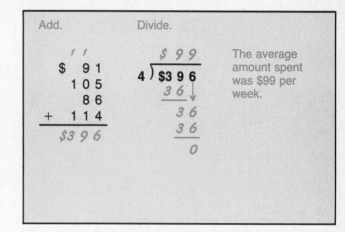

The average amount spent was $99 per week.

Solve.

Do Your Work Here

1. Clarice bowls on the company bowling team. Last week Clarice's scores were 178, 164, and 188. This week her scores were 182, 198, and 212. For those two weeks, what was her bowling average? ___187___

```
 4 3
 1 7 8        1 8 7
 1 6 4     6 ) 1,1 2 2
 1 8 8         6
 1 8 2         5 2
 1 9 8         4 8
+2 1 2          4 2
 _____         4 2
 1,1 2 2          0
```

2. Frank weighs 162 pounds, Thomas weighs 168 pounds, and Emmett weighs 165 pounds. What is their average weight? _____

3. On an auto trip, we drove 50 miles the first hour, 52 miles the second hour, and 54 miles the third hour. How many miles did we average per hour? _____

4. An airplane flew 440 miles per hour with the wind and 380 miles per hour against the wind. What was the average speed for the two hours? _____

5. Darleen Martin works six days a week. Last week she earned $38, $42, $41, $37, $40, and $42. How much did she average per day? _____

6. The linemen on the Ohio State football team weigh 230, 198, 208, 228, 188, 232, and 228 pounds. What is the average weight per player? _____

Estimating Products and Quotients

To estimate products and quotients, round each number in the problem. Then multiply or divide the rounded numbers. Accurate estimating will help you decide if an answer is reasonable.

Estimate: 25 × 211

Round each number to the nearest ten. Multiply.

$$211 \rightarrow 210$$
$$\times\ 25 \rightarrow \times\ \ 30$$
$$6{,}300$$

Estimate: 928 ÷ 29

Round each number to the nearest ten. Divide.

$$928 \rightarrow 930$$
$$29 \rightarrow 30$$

```
        31
   30)930
      90
      30
      30
       0
```

Estimate each product. Round each number to the nearest ten. Multiply.

1.
$$\begin{array}{r} 43 \rightarrow\ 40 \\ \times 82 \rightarrow \times 80 \\ \hline 3{,}200 \end{array}$$
$$\begin{array}{r} 27 \rightarrow \\ \times 14 \rightarrow \\ \hline \end{array}$$
$$\begin{array}{r} 58 \rightarrow \\ \times 33 \rightarrow \\ \hline \end{array}$$
$$\begin{array}{r} 76 \rightarrow \\ \times 65 \rightarrow \\ \hline \end{array}$$

2.
$$\begin{array}{r} 83 \rightarrow \\ \times 26 \rightarrow \\ \hline \end{array}$$
$$\begin{array}{r} 77 \rightarrow \\ \times 55 \rightarrow \\ \hline \end{array}$$
$$\begin{array}{r} 49 \rightarrow \\ \times 37 \rightarrow \\ \hline \end{array}$$
$$\begin{array}{r} 31 \rightarrow \\ \times 94 \rightarrow \\ \hline \end{array}$$

3.
$$\begin{array}{r} 193 \rightarrow \\ \times\ \ 41 \rightarrow \\ \hline \end{array}$$
$$\begin{array}{r} 788 \rightarrow \\ \times\ \ 29 \rightarrow \\ \hline \end{array}$$
$$\begin{array}{r} 572 \rightarrow \\ \times\ \ 15 \rightarrow \\ \hline \end{array}$$
$$\begin{array}{r} 299 \rightarrow \\ \times\ \ 38 \rightarrow \\ \hline \end{array}$$

Estimate each quotient. Round each number to the nearest ten. Divide.

4. $26\overline{)598} \rightarrow 30\overline{)600}^{\,20}$ $38\overline{)802}$ $19\overline{)581}$

5. $12\overline{)664}$ $34\overline{)904}$ $42\overline{)644}$

6. $61\overline{)355}$ $52\overline{)999}$ $55\overline{)777}$

Checking Up

Divide.

1. $6\overline{)72}$ $\quad$ $4\overline{)96}$ $\quad$ $9\overline{)139}$ $\quad$ $7\overline{)154}$ $\quad$ $5\overline{)575}$ $\quad$ $2\overline{)338}$

2. $3\overline{)94}$ $\quad$ $2\overline{)87}$ $\quad$ $5\overline{)128}$ $\quad$ $9\overline{)159}$ $\quad$ $4\overline{)251}$ $\quad$ $8\overline{)255}$

3. $21\overline{)2,395}$ $\quad$ $40\overline{)4,440}$ $\quad$ $32\overline{)6,771}$ $\quad$ $211\overline{)11,816}$ $\quad$ $224\overline{)21,959}$

4. $32\overline{)19,238}$ $\quad$ $18\overline{)36,000}$ $\quad$ $514\overline{)54,484}$ $\quad$ $213\overline{)86,480}$

Estimate each product or quotient. Round each number to the nearest ten.

5. $\begin{array}{r} 92 \\ \times\,11 \\ \hline \end{array}$ $\qquad$ $\begin{array}{r} 196 \\ \times\ \ 31 \\ \hline \end{array}$ $\qquad$ $59\overline{)362}$ $\qquad$ $47\overline{)199}$

Solve.

6. The Harringtons took a vacation. They traveled 330 miles by train, 420 miles by car, and 310 miles by bus. How many miles did they travel in all? If their trip took 10 days, how many miles a day did they average?

   Answer _____

7. Thirty-five tickets to the Big River Amusement Park cost $175. How much does one ticket cost?

   Answer _____

Applying Your Skills

Solve.

1. An orange grower shipped 240 boxes of oranges. Each box contained 112 oranges. How many oranges were sent? _____

2. A movie theater sold out all three showings of a movie. If they sold 2,796 tickets and there were 3 shows, how many people attended each show? _____

3. An airplane travels at an average rate of 310 miles per hour. When it has 430 hours of flying time, how many miles has it flown? _____

4. A store owner bought 60 shirts for $420. How much did she pay per shirt? _____

5. Mars takes about 687 of our days to complete an orbit around the sun. How many earth days would it take to complete 4 orbits around the sun? _____

6. Clovis took 132 hours to sew 11 costumes. If it took the same amount of time to sew each costume, how long did it take to sew one costume? _____

7. An apple grower expects to produce 13,340 apples. If a crate holds 92 apples, how many crates will the grower need? _____

8. A carload of grain weighs 75,000 pounds. If a bushel weighs 60 pounds, how many bushels of grain did the carload contain? _____

9. Juan earns $642 per week. How much will he have earned after 21 weeks? _____

10. Yolanda averages 1,046 miles per week in her job as a consultant. How many miles does she travel in 12 weeks? _____

11. Wong borrowed $3,672 to buy a used car. If he pays the loan back in 36 equal payments, how much is each payment? _____

Add.

1.

936 +377	2,507 +1,530	16,400 + 3,600	5,067 2,049 + 763	2,316 4,219 +1,657	305 13,500 +12,050

Subtract.

2.

897 −276	602 − 27	7,342 − 759	1,944 − 993	3,609 −2,645	2,000 −1,409

Multiply.

3.

434 × 2	1,202 × 14	1,096 × 4	99 ×98	3,247 × 29	3,432 × 212	500 × 50

Divide.

4. $6\overline{)9}$ $9\overline{)171}$ $6\overline{)2,057}$ $36\overline{)1,478}$ $307\overline{)71,838}$ $15\overline{)10,570}$

Write the following in words.

5. 12,313 is written _____ thousand, _____ hundred _____ .

Find the average of the following numbers.

6. 25, 30, 45, 35, 50   _____

Estimate. Round to the nearest hundred.

7.

267 +389	2,365 − 295

Estimate. Round to the nearest ten.

8.

161 × 12	$92\overline{)806}$

Solve.

9. A video store ordered 59 mystery tapes, 72 children's tapes, and 21 comedy tapes. How many tapes were ordered in all?

Answer _____

10. Videotapes are shipped 12 in a box. If a store orders 36 boxes of tapes, how many tapes will be shipped to the store?

Answer _____

Unit 2   Fractions

The Meaning of Fractions

A **fraction** is a part of something. A foot is a fraction of a yard. Specifically, a foot is $\frac{1}{3}$ of a yard. (There are three feet in a whole yard.) We use fractions to tell how much a part is of a whole. Fractions have two parts: a **denominator** and a **numerator.** The bottom number is the denominator. It tells how many parts it takes to make the whole. The top number is the numerator. It tells how many equal parts you have.

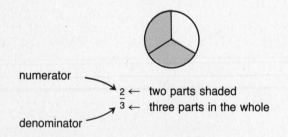

numerator
2 ← two parts shaded
3 ← three parts in the whole
denominator

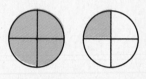

5 ← five parts shaded
4 ← four parts in each whole

Use the drawing at the right to answer these questions.

1. Into how many equal pieces has the rectangle been divided? _____4_____

2. How many parts have been shaded? _____

3. Write the fraction which tells what part is shaded. _____

4. Write the fraction which tells which part is not shaded. _____

5. The rectangle has been divided into fourths. How many fourths equal a whole? _____

6. How many fourths equal one-half? _____

Write the fraction which tells what part is shaded.

7.  $\frac{2}{3}$ _____   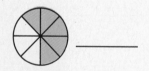 _____   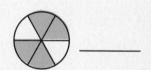 _____

8. 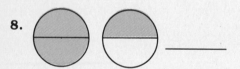 _____    _____   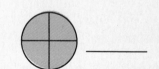 _____

9. 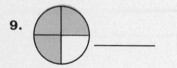 _____    _____   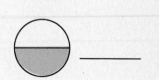 _____

40

Classifying Fractions—Proper, Improper, and Mixed Numbers

There are three forms of fractions—**proper, improper,** and **mixed numbers.** A proper fraction is one in which the numerator is smaller than the denominator. The value of a proper fraction is always less than one. An improper fraction is one in which the numerator is equal to or larger than the denominator. The value of an improper fraction is either equal to one or more than one. A mixed number has a whole number part and a fractional part. The value of a mixed number is always more than one.

EXAMPLES

Proper Fraction	Improper Fractions	Mixed Number
$\frac{4}{5} \rightarrow \dfrac{\text{smaller}}{\text{larger}}$	$\frac{5}{4} \rightarrow \dfrac{\text{larger}}{\text{smaller}}$ $\frac{6}{6} \rightarrow$ equal	$3\frac{2}{5} \rightarrow$ whole number + fraction

Describe the following fractions by writing proper, improper, or mixed number.

1. $\frac{12}{11}$ _____ $\frac{7}{12}$ _____ $\frac{4}{4}$ _____

2. $5\frac{3}{7}$ _____ $\frac{2}{3}$ _____ $\frac{115}{127}$ _____

3. $\frac{3}{5}$ _____ $\frac{5}{3}$ _____ $2\frac{3}{4}$ _____

4. $\frac{10}{10}$ _____ $\frac{9}{20}$ _____ $\frac{1}{2}$ _____

5. $6\frac{1}{4}$ _____ $\frac{7}{5}$ _____ $\frac{4}{3}$ _____

Circle the proper fractions. Put an x on the improper fractions. Do not do anything to mixed numbers.

6. $3\frac{1}{8}$ $\frac{6}{9}$ $\frac{12}{6}$ $1\frac{1}{2}$ $\frac{5}{8}$ $20\frac{1}{2}$ $\frac{13}{4}$ $\frac{8}{8}$

7. $\frac{4}{3}$ $1\frac{3}{4}$ $\frac{1}{8}$ $\frac{9}{7}$ $3\frac{6}{7}$ $\frac{5}{5}$ $10\frac{9}{13}$ $7\frac{4}{5}$

8. $\frac{6}{8}$ $\frac{18}{15}$ $\frac{6}{5}$ $3\frac{1}{2}$ $\frac{9}{12}$ $\frac{1}{8}$ $\frac{20}{21}$ $\frac{1}{9}$

9. $6\frac{1}{3}$ $\frac{3}{7}$ $\frac{15}{2}$ $\frac{6}{2}$ $9\frac{3}{8}$ $1\frac{1}{11}$ $\frac{16}{15}$ $\frac{1}{13}$

10. $\frac{9}{9}$ $2\frac{1}{4}$ $\frac{4}{4}$ $\frac{5}{6}$ $\frac{10}{6}$ $\frac{12}{7}$ $5\frac{7}{8}$ $\frac{3}{12}$

Equivalent Fractions

Different fractions may be used to name the same number. Look at the shapes below. Notice that the same amount of each one is shaded, but the shaded portion of each shape is named with a different fraction.

$\frac{1}{2}$ $\frac{2}{4}$ $\frac{3}{6}$ $\frac{4}{8}$ $\frac{5}{10}$

Two numbers that name the same number are called **equivalent fractions.** For example, $\frac{1}{2}$, $\frac{2}{4}$, $\frac{3}{6}$, $\frac{4}{8}$, and $\frac{5}{10}$ are equivalent fractions because they all name the same number, one-half.

Look at each picture. Write the fractions which name the shaded portion of each shape. Then circle the phrase which correctly expresses the relationship between the fractions.

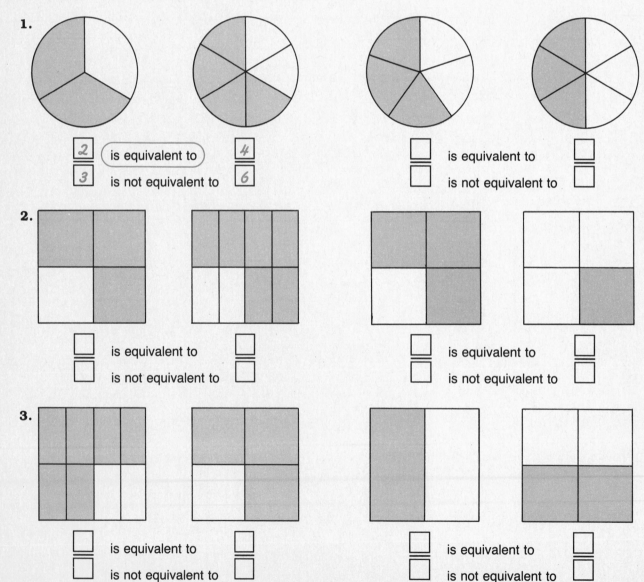

1. $\frac{2}{3}$ (is equivalent to) $\frac{4}{6}$
is not equivalent to

is equivalent to
is not equivalent to

2. is equivalent to
is not equivalent to

is equivalent to
is not equivalent to

3. is equivalent to
is not equivalent to

is equivalent to
is not equivalent to

Changing and Reducing Proper and Improper Fractions

Any fraction may be changed to an equivalent fraction by multiplying or dividing both its numerator and denominator by the **same** number. In addition and subtraction it is often necessary to raise fractions to higher terms by multiplying the numerator and denominator by the same number. Fractions may also be reduced or simplified by dividing the numerator and denominator by the same number. A fraction is reduced to its **lowest terms** when there is no number other than 1 which will divide evenly into both the numerator and denominator. The answers to all addition, subtraction, multiplication, and division problems **must** be reduced to lowest terms (simplified) or your answers will be incorrect.

To change an improper fraction to simplest form, divide the denominator into the numerator for a whole-number quotient. The remainder after dividing becomes the numerator of the fraction. The divisor becomes the denominator of the fraction.

Raise $\frac{1}{2}$ to fourths.

Multiply both numerator and denominator by 2.
$\frac{1 \times 2}{2 \times 2} = \frac{2}{4}$

Reduce $\frac{10}{15}$ to lowest terms.

Divide both numerator and denominator by 5.
$\frac{10 \div 5}{15 \div 5} = \frac{2}{3}$

Simplify $\frac{14}{4}$.

First divide 4 into 14.
$4 \overline{)14} \quad \frac{3}{} \rightarrow 3\frac{2}{4} = 3\frac{1}{2}$ $\quad\underline{12}$ $\quad 2$ Write the remainder as a fraction.

Raise each fraction to higher terms as indicated.

1. $\frac{2}{3} = \frac{6}{9}$ $\qquad$ $\frac{1}{2} = \frac{}{12}$ $\qquad$ $\frac{1}{4} = \frac{}{12}$ $\qquad$ $\frac{1}{4} = \frac{}{8}$ $\qquad$ $\frac{5}{6} = \frac{}{12}$

2. $\frac{1}{2} = \frac{}{32}$ $\qquad$ $\frac{5}{8} = \frac{}{32}$ $\qquad$ $\frac{3}{4} = \frac{}{32}$ $\qquad$ $\frac{3}{4} = \frac{}{16}$ $\qquad$ $\frac{5}{8} = \frac{}{16}$

Reduce each fraction to lowest terms.

3. $\frac{8}{12} = \frac{2}{3}$ $\qquad$ $\frac{8}{16} =$ $\qquad$ $\frac{6}{8} =$ $\qquad$ $\frac{3}{9} =$ $\qquad$ $\frac{9}{12} =$

4. $\frac{4}{8} =$ $\qquad$ $\frac{8}{12} =$ $\qquad$ $\frac{15}{30} =$ $\qquad$ $\frac{20}{25} =$ $\qquad$ $\frac{4}{8} =$

Simplify the following improper fractions. Be sure to write answers in lowest terms.

5. $\frac{7}{2} = 3\frac{1}{2}$ $\qquad$ $\frac{6}{5} =$ $\qquad$ $\frac{13}{4} =$ $\qquad$ $\frac{18}{4} =$ $\qquad$ $\frac{21}{7} =$

6. $\frac{52}{3} =$ $\qquad$ $\frac{20}{5} =$ $\qquad$ $\frac{39}{6} =$ $\qquad$ $\frac{100}{3} =$ $\qquad$ $\frac{9}{8} =$

Comparing Fractions

If two fractions have the same denominator but different numerators, the larger numerator tells you that fraction is larger than the other fraction. When two fractions have different denominators, you need to rename the fractions as fractions with common denominators before comparing the numerators.

$\frac{1}{8}$ $\frac{2}{8}$ $\frac{3}{8}$ $\frac{4}{8}$ $\frac{5}{8}$ $\frac{6}{8}$ $\frac{7}{8}$ $\frac{8}{8}$

Put each group of fractions in order from smallest to largest.

1. $\frac{3}{5}$, $\frac{1}{5}$, $\frac{5}{5}$, $\frac{0}{5}$, $\frac{2}{5}$, $\frac{4}{5}$ _____ $\frac{0}{5}$, $\frac{1}{5}$, $\frac{2}{5}$, $\frac{3}{5}$, $\frac{4}{5}$, $\frac{5}{5}$

2. $\frac{5}{8}$, $\frac{7}{8}$, $\frac{3}{8}$, $\frac{1}{8}$, $\frac{9}{8}$, $\frac{11}{8}$ _____

3. $\frac{2}{4}$, $\frac{5}{4}$, $\frac{0}{4}$, $\frac{1}{4}$, $\frac{6}{4}$, $\frac{3}{4}$ _____

4. $\frac{1}{6}$, $\frac{8}{6}$, $\frac{3}{6}$, $\frac{7}{6}$, $\frac{4}{6}$, $\frac{9}{6}$ _____

5. $\frac{5}{11}$, $\frac{1}{11}$, $\frac{2}{11}$, $\frac{9}{11}$, $\frac{0}{11}$, $\frac{11}{11}$ _____

Change the fractions in each pair to common denominators. Then circle the larger fraction.

6. $\frac{1}{4}$ $\left(\frac{5}{16}\right)$ $\frac{7}{16}$ $\frac{3}{8}$ $\frac{1}{2}$ $\frac{9}{16}$ $\frac{11}{16}$ $\frac{19}{32}$

    $\frac{4}{16}$ ← compare → $\frac{5}{16}$

7. $\frac{5}{8}$ $\frac{3}{4}$ $\frac{25}{32}$ $\frac{13}{16}$ $\frac{7}{8}$ $\frac{3}{4}$ $\frac{3}{8}$ $\frac{5}{16}$

Change the fractions to common denominators.
Write the fractions in order from least to greatest.

8. $\frac{7}{8}$ $\frac{3}{4}$ $\frac{3}{8}$ _____ $\frac{3}{8}$ $\frac{6}{8}$ $\frac{7}{8}$ $\frac{1}{3}$ $\frac{3}{6}$ $\frac{1}{6}$ _____

9. $\frac{2}{3}$ $\frac{2}{6}$ $\frac{5}{6}$ _____ $\frac{1}{4}$ $\frac{2}{4}$ $\frac{3}{8}$ _____

10. $\frac{3}{8}$ $\frac{1}{8}$ $\frac{3}{4}$ _____ $\frac{2}{3}$ $\frac{1}{3}$ $\frac{3}{6}$ _____

Adding Fractions with Like Denominators

To add fractions with like denominators, add the numerators. Use the same denominator. Simplify the answer. Remember, to simplify an improper fraction, write it as a whole number or mixed number. To simplify a proper fraction, write it in lowest terms.

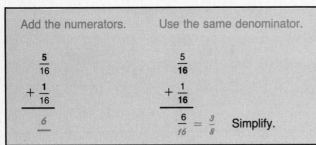

Find: $\frac{5}{16} + \frac{1}{16}$

Add the numerators.   Use the same denominator.

$$\frac{5}{16} + \frac{1}{16} = \frac{6}{}$$

$$\frac{5}{16} + \frac{1}{16} = \frac{6}{16} = \frac{3}{8} \quad \text{Simplify.}$$

Find: $\frac{7}{8} + \frac{3}{8}$

Add the numerators.   Use the same denominator.

$$\frac{7}{8} + \frac{3}{8} = \frac{10}{}$$

$$\frac{7}{8} + \frac{3}{8} = \frac{10}{8} = 1\frac{2}{8} = 1\frac{1}{4} \quad \text{Simplify.}$$

Add the following fractions. Simplify by writing the sums in lowest terms.

1.

| $\frac{4}{8} + \frac{1}{8} = \frac{5}{8}$ | $\frac{1}{5} + \frac{2}{5}$ | $\frac{3}{9} + \frac{1}{9}$ | $\frac{2}{4} + \frac{1}{4}$ | $\frac{6}{10} + \frac{1}{10}$ | $\frac{3}{6} + \frac{2}{6}$ | $\frac{1}{3} + \frac{1}{3}$ | $\frac{2}{8} + \frac{3}{8}$ |

2.

| $\frac{4}{12} + \frac{7}{12} = \frac{11}{12}$ | $\frac{5}{10} + \frac{4}{10}$ | $\frac{6}{15} + \frac{5}{15}$ | $\frac{6}{20} + \frac{3}{20}$ | $\frac{3}{9} + \frac{5}{9}$ | $\frac{4}{6} + \frac{1}{6}$ | $\frac{4}{8} + \frac{3}{8}$ | $\frac{2}{12} + \frac{5}{12}$ |

3.

| $\frac{6}{12} + \frac{3}{12} = \frac{9}{12} = \frac{3}{4}$ | $\frac{3}{20} + \frac{5}{20}$ | $\frac{3}{9} + \frac{2}{9}$ | $\frac{3}{16} + \frac{4}{16}$ | $\frac{3}{8} + \frac{1}{8}$ | $\frac{3}{10} + \frac{1}{10}$ | $\frac{2}{6} + \frac{1}{6}$ | $\frac{2}{12} + \frac{2}{12}$ |

4.

| $\frac{1}{3} + \frac{5}{3} + \frac{2}{3} = \frac{8}{3} = 2\frac{2}{3}$ | $\frac{5}{2} + \frac{3}{2} + \frac{1}{2}$ | $\frac{6}{12} + \frac{4}{12} + \frac{5}{12}$ | $\frac{3}{5} + \frac{1}{5} + \frac{2}{5}$ | $\frac{3}{6} + \frac{5}{6} + \frac{2}{6}$ | $\frac{1}{3} + \frac{2}{3} + \frac{1}{3}$ | $\frac{4}{10} + \frac{6}{10} + \frac{3}{10}$ | $\frac{3}{8} + \frac{5}{8} + \frac{2}{8}$ |

5.

| $\frac{8}{16} + \frac{5}{16} + \frac{7}{16}$ | $\frac{7}{12} + \frac{9}{12} + \frac{5}{12}$ | $\frac{9}{10} + \frac{7}{10} + \frac{3}{10}$ | $\frac{3}{8} + \frac{7}{8} + \frac{5}{8}$ | $\frac{7}{5} + \frac{4}{5} + \frac{3}{5}$ | $\frac{6}{4} + \frac{5}{4} + \frac{3}{4}$ | $\frac{4}{3} + \frac{2}{3} + \frac{5}{3}$ | $\frac{5}{2} + \frac{4}{2} + \frac{7}{2}$ |

Adding Fractions with Unlike Denominators

To add fractions with unlike denominators, first rewrite the fractions as equivalent fractions with like or common denominators. Then add the numerators. Use the same denominator.

Find: $\frac{1}{4} + \frac{8}{16}$

Write equivalent fractions with 16 as the common denominator.

$$\frac{1}{4} = \frac{4}{16}$$

Remember,

$$+ \frac{8}{16} = \frac{8}{16}$$

$$\frac{1}{4} = \frac{1 \times 4}{4 \times 4} = \frac{4}{16}$$

Add the numerators. Use the same denominator.

$$\frac{1}{4} = \frac{4}{16}$$

$$+ \frac{8}{16} = \frac{8}{16}$$

$$\frac{12}{16} = \frac{3}{4} \quad \text{Simplify.}$$

Add the following fractions. Simplify by writing the answer in lowest terms.

1. $\frac{1}{12} = \frac{1}{12}$ $\frac{1}{6}$ $\frac{1}{2}$ $\frac{1}{8}$ $\frac{1}{8}$ $\frac{1}{4}$

$+ \frac{1}{4} = \frac{3}{12}$ $+ \frac{1}{3}$ $+ \frac{1}{6}$ $+ \frac{1}{2}$ $+ \frac{1}{4}$ $+ \frac{1}{2}$

$\frac{4}{12} = \frac{1}{3}$

2. $\frac{2}{12}$ $\frac{3}{8}$ $\frac{3}{6}$ $\frac{2}{6}$ $\frac{3}{8}$ $\frac{1}{8}$

$+ \frac{1}{4}$ $+ \frac{1}{4}$ $+ \frac{1}{3}$ $+ \frac{1}{2}$ $+ \frac{1}{2}$ $+ \frac{3}{4}$

3. $\frac{1}{6}$ $\frac{2}{10}$ $\frac{2}{9}$ $\frac{4}{8}$ $\frac{3}{10}$

$+ \frac{1}{3}$ $+ \frac{2}{5}$ $+ \frac{2}{3}$ $+ \frac{1}{4}$ $+ \frac{1}{5}$

4. $\frac{1}{2} = \frac{5}{10}$ $\frac{1}{2}$ $\frac{3}{4}$ $\frac{1}{2}$ $\frac{5}{8}$

$+ \frac{7}{10} = \frac{7}{10}$ $+ \frac{4}{6}$ $+ \frac{5}{8}$ $+ \frac{3}{4}$ $+ \frac{1}{2}$

$\frac{12}{10} = 1\frac{2}{10} = 1\frac{1}{5}$

5. $\frac{1}{2}$ $\frac{7}{16}$ $\frac{7}{10}$ $\frac{3}{4}$ $\frac{7}{15}$

$+ \frac{7}{12}$ $+ \frac{3}{4}$ $+ \frac{4}{5}$ $+ \frac{9}{12}$ $+ \frac{4}{5}$

6. $\frac{5}{6} = \frac{5}{6}$ $\frac{7}{9}$ $\frac{5}{9}$ $\frac{2}{3}$ $\frac{2}{3}$

$+ \frac{1}{2} = \frac{3}{6}$ $+ \frac{2}{3}$ $+ \frac{1}{3}$ $+ \frac{4}{21}$ $+ \frac{7}{15}$

$\frac{8}{6} = 1\frac{2}{6} = 1\frac{1}{3}$

Using the LCD

To add or subtract fractions, you might need to find the *lowest* common denominator (LCD) of the fractions. The lowest common denominator is the smallest number that both denominators will divide into evenly. Then you can use the LCD to write equivalent fractions.

Use the LCD to write equivalent fractions for $\frac{2}{3}$ and $\frac{1}{4}$.

List several multiples of each denominator.	The LCD is the smallest number that appears on both lists.	Write equivalent fractions.
Multiples of 3: 3 6 9 12 15 Multiples of 4: 4 8 12 16 20	The LCD of $\frac{2}{3}$ and $\frac{1}{4}$ is 12. The LCD is the smallest number that both 3 and 4 will divide into evenly.	$\frac{2}{3} = \frac{2 \times 4}{3 \times 4} = \frac{8}{12}$ $\frac{1}{4} = \frac{1 \times 3}{4 \times 3} = \frac{3}{12}$

Find: $\frac{2}{3} + \frac{1}{5}$

Write equivalent fractions with like denominators. Use the LCD.	Add the numerators. Use the same denominator.
$\frac{2}{3} = \frac{2 \times 5}{3 \times 5} = \frac{10}{15}$ $+ \frac{1}{5} = \frac{1 \times 3}{5 \times 3} = \frac{3}{15}$	$\frac{2}{3} = \frac{10}{15}$ $+ \frac{1}{5} = \frac{3}{15}$ $\overline{\phantom{+\frac{1}{5}}\frac{13}{15}}$

Use the LCD to write equivalent fractions for each pair of fractions below.

1. $\frac{3}{4} = \frac{3 \times 3}{4 \times 3} = \frac{9}{12}$ $\quad\Big|\quad$ $\frac{1}{3} =$ $\qquad$ $\frac{2}{5} =$ $\qquad$ $\frac{5}{6} =$ $\qquad$ $\frac{3}{4} =$

$\frac{1}{6} = \frac{1 \times 2}{6 \times 2} = \frac{2}{12}$ $\quad\Big|\quad$ $\frac{4}{5} =$ $\qquad$ $\frac{1}{4} =$ $\qquad$ $\frac{7}{8} =$ $\qquad$ $\frac{1}{3} =$

Add. Simplify.

2. $\quad \frac{1}{2} = \frac{9}{18}$ $\qquad\quad \frac{4}{5}$ $\qquad\qquad\quad \frac{1}{3}$ $\qquad\qquad\quad \frac{3}{7}$ $\qquad\qquad\quad \frac{2}{3}$

$\quad + \frac{7}{9} = \frac{14}{18}$ $\qquad + \frac{2}{3}$ $\qquad\qquad + \frac{3}{10}$ $\qquad\qquad + \frac{1}{2}$ $\qquad\qquad + \frac{4}{9}$

$\qquad\quad \frac{23}{18} = 1\frac{5}{18}$

3. $\quad \frac{2}{3}$ $\qquad\qquad\quad \frac{5}{6}$ $\qquad\qquad\quad \frac{3}{7}$ $\qquad\qquad\quad \frac{1}{2}$ $\qquad\qquad\quad \frac{3}{5}$

$\quad + \frac{3}{4}$ $\qquad\qquad + \frac{3}{5}$ $\qquad\qquad + \frac{1}{4}$ $\qquad\qquad + \frac{2}{8}$ $\qquad\qquad + \frac{1}{4}$

Adding Fractions and Reducing

Check for unlike denominators. Add the fractions and simplify the answers.

1.
$\frac{2}{4}$
$+\frac{3}{4}$
$\frac{5}{4} = 1\frac{1}{4}$

$\frac{2}{8}$
$+\frac{3}{8}$

$\frac{1}{3}$
$+\frac{2}{3}$

$\frac{3}{2}$
$+\frac{1}{2}$

$\frac{7}{5}$
$+\frac{2}{5}$

$\frac{4}{6}$
$+\frac{5}{6}$

2.
$\frac{7}{6}$
$+\frac{7}{6}$
$\frac{14}{6} = 2\frac{2}{6} = 2\frac{1}{3}$

$\frac{9}{10}$
$+\frac{9}{10}$

$\frac{7}{8}$
$+\frac{5}{8}$

$\frac{5}{4}$
$+\frac{5}{4}$

$\frac{12}{16}$
$+\frac{17}{16}$

$\frac{11}{12}$
$+\frac{9}{12}$

3.
$\frac{4}{5} = \frac{8}{10}$
$\frac{8}{10} = \frac{8}{10}$
$+\frac{6}{10} = \frac{6}{10}$
$\frac{22}{10} = 2\frac{2}{10} = 2\frac{1}{5}$

$\frac{2}{8}$
$\frac{5}{8}$
$+\frac{7}{8}$

$\frac{4}{5}$
$\frac{10}{15}$
$+\frac{9}{15}$

$\frac{10}{20}$
$\frac{8}{20}$
$+\frac{12}{20}$

$\frac{10}{16}$
$\frac{18}{32}$
$+\frac{12}{32}$

$\frac{2}{9}$
$\frac{4}{9}$
$+\frac{5}{9}$

4.
$\frac{6}{4}$
$\frac{7}{4}$
$+\frac{3}{4}$
$\frac{16}{4} = 4$

$\frac{8}{16}$
$\frac{15}{16}$
$+\frac{9}{16}$

$\frac{9}{5}$
$\frac{2}{5}$
$+\frac{4}{5}$

$\frac{6}{8}$
$\frac{7}{8}$
$+\frac{3}{8}$

$\frac{4}{2}$
$\frac{3}{2}$
$+\frac{1}{2}$

$\frac{1}{3}$
$\frac{6}{3}$
$+\frac{2}{3}$

5.
$\frac{3}{4} = \frac{9}{12}$
$\frac{2}{3} = \frac{8}{12}$
$+\frac{1}{2} = \frac{6}{12}$
$\frac{23}{12} = 1\frac{11}{12}$

$\frac{4}{6}$
$\frac{9}{6}$
$+\frac{5}{6}$

$\frac{7}{8}$
$\frac{1}{4}$
$+\frac{1}{2}$

$\frac{15}{20}$
$\frac{16}{20}$
$+\frac{9}{20}$

$\frac{1}{3}$
$\frac{5}{6}$
$+\frac{1}{4}$

$\frac{4}{5}$
$\frac{9}{5}$
$+\frac{7}{5}$

6.
$\frac{8}{12}$
$\frac{8}{12}$
$+\frac{9}{12}$

$\frac{4}{5}$
$\frac{12}{10}$
$+\frac{9}{10}$

$\frac{2}{4}$
$\frac{11}{6}$
$+\frac{5}{6}$

$\frac{3}{15}$
$\frac{16}{5}$
$+\frac{4}{5}$

$\frac{3}{8}$
$\frac{15}{8}$
$+\frac{7}{8}$

$\frac{7}{4}$
$\frac{11}{4}$
$+\frac{3}{4}$

7.
$\frac{4}{5}$
$\frac{3}{5}$
$\frac{2}{5}$
$+\frac{3}{5}$

$\frac{5}{7}$
$\frac{4}{7}$
$\frac{6}{7}$
$+\frac{3}{7}$

$\frac{9}{10}$
$\frac{7}{10}$
$\frac{3}{5}$
$+\frac{1}{2}$

$\frac{9}{12}$
$\frac{7}{12}$
$\frac{5}{12}$
$+\frac{3}{12}$

$\frac{1}{2}$
$\frac{1}{4}$
$\frac{2}{8}$
$+\frac{5}{8}$

$\frac{17}{20}$
$\frac{9}{20}$
$\frac{19}{20}$
$+\frac{3}{20}$

48

Adding Mixed Numbers and Fractions

When adding mixed numbers and fractions, first check for unlike denominators. Write mixed numbers and fractions as equivalent fractions with like denominators. Add the fractions. Then add the whole numbers. You may need to regroup to simplify your answer.

Find: $3\frac{5}{6} + 1\frac{3}{4}$

Write the fractions with like denominators.	Add the fractions.	Add the whole numbers.	$4\frac{19}{12}$ is a mixed number with an improper fraction. Regroup to simplify.
$3\frac{5}{6} = 3\frac{10}{12}$	$3\frac{5}{6} = 3\frac{10}{12}$	$3\frac{5}{6} = 3\frac{10}{12}$	
$+1\frac{3}{4} = 1\frac{9}{12}$	$+1\frac{3}{4} = 1\frac{9}{12}$	$+1\frac{3}{4} = 1\frac{9}{12}$	$4\frac{19}{12} = 4 + 1\frac{7}{12} = 5\frac{7}{12}$
	$\frac{19}{12}$	$4\frac{19}{12}$	

Add. Simplify.

1. $5\frac{1}{3} = 5\frac{2}{6}$ $\frac{2}{5}$ $3\frac{6}{7}$ $\frac{5}{6}$ $8\frac{1}{2}$

    $+\ \frac{5}{6} = \ \frac{5}{6}$ $+2\frac{7}{10}$ $+\ \frac{3}{14}$ $+9\frac{1}{2}$ $+\ \frac{3}{4}$

       $5\frac{7}{6} = 6\frac{1}{6}$

2. $11\frac{4}{9} = 11\frac{4}{9}$ $2\frac{1}{4}$ $11\frac{1}{2}$ $6\frac{4}{15}$ $7\frac{1}{4}$

    $+\ 4\frac{1}{3} = \ 4\frac{3}{9}$ $+3\frac{1}{8}$ $+\ 4\frac{3}{8}$ $+3\frac{1}{3}$ $+12\frac{3}{8}$

       $15\frac{7}{9}$

3. $6\frac{1}{2} = 6\frac{7}{14}$ $1\frac{3}{4}$ $3\frac{1}{2}$ $12\frac{2}{5}$ $9\frac{2}{3}$

    $+2\frac{4}{7} = 2\frac{8}{14}$ $+5\frac{1}{3}$ $+6\frac{4}{5}$ $+\ 7\frac{3}{4}$ $+10\frac{5}{8}$

       $8\frac{15}{14} = 9\frac{1}{14}$

4. $\frac{1}{10}$ $3\frac{2}{3}$ $\frac{3}{8}$ $8\frac{3}{5}$ $\frac{3}{4}$

    $+2\frac{1}{2}$ $+\ \frac{3}{4}$ $+4\frac{1}{6}$ $+13\frac{1}{2}$ $+7\frac{5}{6}$

5. $29\frac{2}{3}$ $16\frac{1}{2}$ $18\frac{3}{10}$ $3\frac{1}{5}$ $16\frac{2}{3}$

    $+\ 7\frac{5}{6}$ $+\ \frac{1}{9}$ $+23\frac{3}{5}$ $+\ \frac{2}{3}$ $+13\frac{5}{9}$

Raise to higher terms.

1. $\frac{1}{3} = \frac{}{12}$ $\frac{1}{2} = \frac{}{8}$ $\frac{1}{4} = \frac{}{16}$ $\frac{1}{4} = \frac{}{12}$ $\frac{1}{3} = \frac{}{9}$ $\frac{1}{2} = \frac{}{16}$

2. $\frac{2}{5} = \frac{}{10}$ $\frac{3}{4} = \frac{}{16}$ $\frac{4}{5} = \frac{}{15}$ $\frac{5}{6} = \frac{}{12}$ $\frac{5}{8} = \frac{}{16}$ $\frac{2}{3} = \frac{}{12}$

Reduce these fractions to lowest terms.

3. $\frac{9}{12} = $ ____ $\frac{8}{12} = $ ____ $\frac{9}{15} = $ ____ $\frac{6}{12} = $ ____ $\frac{6}{10} = $ ____ $\frac{6}{9} = $ ____

4. $\frac{15}{25} = $ ____ $\frac{6}{8} = $ ____ $\frac{12}{16} = $ ____ $\frac{4}{16} = $ ____ $\frac{2}{8} = $ ____ $\frac{5}{15} = $ ____

Simplify these fractions.

5. $\frac{18}{15} = $ ____ $\frac{14}{6} = $ ____ $\frac{10}{4} = $ ____ $\frac{9}{6} = $ ____ $\frac{12}{8} = $ ____ $\frac{13}{5} = $ ____

6. $\frac{15}{5} = $ ____ $\frac{8}{2} = $ ____ $\frac{9}{3} = $ ____ $\frac{15}{8} = $ ____ $\frac{7}{2} = $ ____ $\frac{9}{4} = $ ____

Add.

7.

$\begin{array}{r} \frac{1}{3} \\ + \frac{2}{3} \\ \hline \end{array}$ $\begin{array}{r} \frac{2}{5} \\ + \frac{3}{5} \\ \hline \end{array}$ $\begin{array}{r} \frac{5}{16} \\ + \frac{7}{16} \\ \hline \end{array}$ $\begin{array}{r} \frac{5}{8} \\ + \frac{1}{8} \\ \hline \end{array}$ $\begin{array}{r} \frac{3}{4} \\ + \frac{1}{4} \\ \hline \end{array}$

8.

$\begin{array}{r} \frac{2}{3} \\ + \frac{4}{5} \\ \hline \end{array}$ $\begin{array}{r} \frac{2}{3} \\ + \frac{1}{2} \\ \hline \end{array}$ $\begin{array}{r} \frac{3}{8} \\ + \frac{1}{2} \\ \hline \end{array}$ $\begin{array}{r} \frac{5}{9} \\ + \frac{1}{3} \\ \hline \end{array}$ $\begin{array}{r} \frac{1}{3} \\ + \frac{3}{4} \\ \hline \end{array}$

9.

$\begin{array}{r} \frac{1}{4} \\ \frac{1}{4} \\ + \frac{3}{4} \\ \hline \end{array}$ $\begin{array}{r} \frac{2}{3} \\ \frac{1}{3} \\ + \frac{2}{5} \\ \hline \end{array}$ $\begin{array}{r} \frac{1}{2} \\ \frac{1}{6} \\ + \frac{5}{6} \\ \hline \end{array}$ $\begin{array}{r} \frac{6}{8} \\ \frac{7}{8} \\ + \frac{1}{4} \\ \hline \end{array}$ $\begin{array}{r} \frac{1}{2} \\ \frac{3}{5} \\ + \frac{4}{5} \\ \hline \end{array}$

10.

$\begin{array}{r} 7\frac{1}{6} \\ + 3\frac{4}{5} \\ \hline \end{array}$ $\begin{array}{r} 16\frac{3}{4} \\ + 15\frac{2}{3} \\ \hline \end{array}$ $\begin{array}{r} 14\frac{7}{8} \\ + 6\frac{1}{2} \\ \hline \end{array}$ $\begin{array}{r} 6\frac{1}{2} \\ + 7\frac{1}{3} \\ \hline \end{array}$ $\begin{array}{r} 19\frac{3}{10} \\ + 1\frac{4}{5} \\ \hline \end{array}$

Solve.

Do Your Work Here

11. One end of a wrench is marked $\frac{3}{16}$ in. The other end is marked $\frac{1}{4}$ in. Which measure is larger?   _____

Subtracting Fractions with Like Denominators

To subtract fractions with like denominators, subtract the numerators. Use the same denominator. Simplify the answer. To simplify an improper fraction, write it as a whole number or mixed number. To simplify a proper fraction, write it in simplest terms.

Find: $\frac{7}{8} - \frac{3}{8}$

Find: $\frac{12}{20} - \frac{6}{20}$

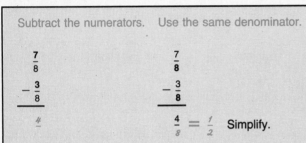

Subtract the numerators.   Use the same denominator.

$\begin{array}{r} \frac{7}{8} \\ -\frac{3}{8} \\ \hline \frac{4}{} \end{array}$

$\begin{array}{r} \frac{7}{8} \\ -\frac{3}{8} \\ \hline \frac{4}{8} \end{array} = \frac{1}{2}$ Simplify.

Subtract the numerators.   Use the same denominator.

$\begin{array}{r} \frac{12}{20} \\ -\frac{6}{20} \\ \hline \frac{6}{} \end{array}$

$\begin{array}{r} \frac{12}{20} \\ -\frac{6}{20} \\ \hline \frac{6}{20} \end{array} = \frac{3}{10}$ Simplify.

Subtract. Simplify by writing the difference in lowest terms.

1. $\frac{9}{10} - \frac{8}{10} = \frac{1}{10}$ $\frac{12}{16} - \frac{5}{16}$ $\frac{11}{12} - \frac{6}{12}$ $\frac{5}{9} - \frac{4}{9}$ $\frac{3}{4} - \frac{2}{4}$ $\frac{5}{6} - \frac{4}{6}$

2. $\frac{15}{16} - \frac{7}{16} = \frac{8}{16} = \frac{1}{2}$ $\frac{9}{10} - \frac{4}{10}$ $\frac{9}{12} - \frac{5}{12}$ $\frac{5}{9} - \frac{2}{9}$ $\frac{5}{6} - \frac{2}{6}$ $\frac{5}{8} - \frac{3}{8}$

3. $\frac{8}{9} - \frac{2}{9}$ $\frac{7}{9} - \frac{4}{9}$ $\frac{5}{6} - \frac{3}{6}$ $\frac{3}{4} - \frac{1}{4}$ $\frac{9}{16} - \frac{5}{16}$ $\frac{7}{8} - \frac{3}{8}$

4. $\frac{17}{32} - \frac{11}{32}$ $\frac{40}{64} - \frac{24}{64}$ $\frac{75}{100} - \frac{25}{100}$ $\frac{12}{20} - \frac{8}{20}$ $\frac{12}{15} - \frac{3}{15}$ $\frac{9}{12} - \frac{5}{12}$

5. $\frac{7}{10} - \frac{2}{10}$ $\frac{5}{6} - \frac{1}{6}$ $\frac{7}{8} - \frac{1}{8}$ $\frac{9}{10} - \frac{7}{10}$ $\frac{7}{10} - \frac{5}{10}$ $\frac{3}{5} - \frac{1}{5}$

6. $\frac{7}{10} - \frac{1}{10}$ $\frac{8}{9} - \frac{5}{9}$ $\frac{9}{10} - \frac{1}{10}$ $\frac{7}{8} - \frac{5}{8}$ $\frac{9}{16} - \frac{3}{16}$ $\frac{7}{8} - \frac{3}{8}$

Subtracting Fractions with Unlike Denominators

To subtract fractions with unlike denominators, first rewrite the fractions as fractions with like denominators. Use the lowest common denominator (LCD). Then subtract and simplify the answer.

Find: $\frac{5}{6} - \frac{3}{8}$

Write equivalent fractions with 24 as the LCD. Both 6 and 8 will divide evenly into 24.

$$\frac{5}{6} = \frac{5 \times 4}{6 \times 4} = \frac{20}{24}$$
$$-\frac{3}{8} = \frac{3 \times 3}{8 \times 3} = \frac{9}{24}$$

Subtract the numerators. Use the same denominator.

$$\frac{5}{6} = \frac{20}{24}$$
$$-\frac{3}{8} = \frac{9}{24}$$
$$\frac{11}{24}$$

Subtract. Simplify where possible.

1.
$\frac{3}{4} = \frac{6}{8}$ $\quad$ $\frac{5}{8}$ $\quad$ $\frac{5}{6}$ $\quad$ $\frac{1}{2}$ $\quad$ $\frac{7}{9}$
$-\frac{3}{8} = \frac{3}{8}$ $\quad$ $-\frac{1}{2}$ $\quad$ $-\frac{1}{3}$ $\quad$ $-\frac{1}{4}$ $\quad$ $-\frac{2}{3}$
$\frac{3}{8}$

2.
$\frac{1}{3}$ $\quad$ $\frac{1}{2}$ $\quad$ $\frac{1}{2}$ $\quad$ $\frac{9}{10}$ $\quad$ $\frac{3}{4}$
$-\frac{1}{6}$ $\quad$ $-\frac{3}{7}$ $\quad$ $-\frac{1}{8}$ $\quad$ $-\frac{1}{4}$ $\quad$ $-\frac{1}{2}$

3.
$\frac{5}{6}$ $\quad$ $\frac{1}{16}$ $\quad$ $\frac{1}{8}$ $\quad$ $\frac{1}{6}$ $\quad$ $\frac{1}{5}$
$-\frac{1}{4}$ $\quad$ $-\frac{1}{32}$ $\quad$ $-\frac{1}{10}$ $\quad$ $-\frac{1}{12}$ $\quad$ $-\frac{1}{6}$

4.
$\frac{17}{20}$ $\quad$ $\frac{9}{10}$ $\quad$ $\frac{11}{16}$ $\quad$ $\frac{8}{9}$ $\quad$ $\frac{15}{32}$
$-\frac{1}{2}$ $\quad$ $-\frac{1}{3}$ $\quad$ $-\frac{1}{4}$ $\quad$ $-\frac{1}{2}$ $\quad$ $-\frac{1}{8}$

5.
$\frac{11}{3} = \frac{22}{6}$ $\quad$ $\frac{15}{8}$ $\quad$ $\frac{16}{6}$ $\quad$ $\frac{20}{5}$ $\quad$ $\frac{12}{10}$
$-\frac{1}{2} = \frac{3}{6}$ $\quad$ $-\frac{1}{4}$ $\quad$ $-\frac{1}{5}$ $\quad$ $-\frac{3}{4}$ $\quad$ $-\frac{1}{5}$
$\frac{19}{6} = 3\frac{1}{6}$

6.
$\frac{11}{10}$ $\quad$ $\frac{16}{7}$ $\quad$ $\frac{11}{8}$ $\quad$ $\frac{13}{6}$ $\quad$ $\frac{12}{9}$
$-\frac{1}{2}$ $\quad$ $-\frac{1}{2}$ $\quad$ $-\frac{1}{4}$ $\quad$ $-\frac{2}{5}$ $\quad$ $-\frac{1}{2}$

Subtracting Fractions from Mixed Numbers

Fractions are subtracted from mixed numbers in much the same way as fractions are subtracted from fractions.

Find: $8\frac{1}{2} - \frac{1}{3}$

Write the mixed numbers with like denominators. Use the LCD.	Subtract the fractions.	Subtract the whole numbers.
$8\frac{1}{2} = 8\frac{3}{6}$ $-\frac{1}{3} = \frac{2}{6}$	$8\frac{1}{2} = 8\frac{3}{6}$ $-\frac{1}{3} = \frac{2}{6}$ $\frac{1}{6}$	$8\frac{1}{2} = 8\frac{3}{6}$ $-\frac{1}{3} = \frac{2}{6}$ $8\frac{1}{6}$

Subtract. Simplify where possible.

1. $7\frac{5}{8} = 7\frac{5}{8}$
   $-\frac{1}{4} = \frac{2}{8}$
   $7\frac{3}{8}$

   $9\frac{5}{8}$
   $-\frac{1}{3}$

   $15\frac{3}{4}$
   $-\frac{1}{2}$

   $12\frac{5}{6}$
   $-\frac{1}{4}$

2. $8\frac{7}{10} = 8\frac{7}{10}$
   $-\frac{1}{2} = \frac{5}{10}$
   $8\frac{2}{10} = 8\frac{1}{5}$

   $16\frac{5}{7}$
   $-\frac{1}{3}$

   $15\frac{2}{3}$
   $-\frac{1}{6}$

   $19\frac{1}{2}$
   $-\frac{1}{5}$

3. $7\frac{1}{2}$
   $-\frac{1}{4}$

   $16\frac{3}{4}$
   $-\frac{1}{3}$

   $11\frac{3}{5}$
   $-\frac{2}{10}$

   $7\frac{1}{2}$
   $-\frac{3}{7}$

4. $10\frac{2}{5}$
   $-\frac{1}{4}$

   $6\frac{2}{3}$
   $-\frac{3}{10}$

   $12\frac{2}{3}$
   $-\frac{1}{4}$

   $8\frac{3}{16}$
   $-\frac{1}{8}$

5. $9\frac{3}{4}$
   $-\frac{1}{6}$

   $6\frac{2}{3}$
   $-\frac{1}{5}$

   $8\frac{3}{4}$
   $-\frac{5}{12}$

   $9\frac{1}{2}$
   $-\frac{1}{5}$

Solve.

6. To make a shelf, Georgia cut $\frac{5}{16}$ inch off a board measuring $38\frac{3}{4}$ inches. How long was the shelf?

   Answer _____

7. Al bought a ham weighing $12\frac{7}{8}$ pounds. The butcher cut off one $\frac{1}{2}$-pound slice. How many pounds were left in the ham?

   Answer _____

Subtracting Mixed Numbers

To subtract mixed numbers, first subtract the fractions. Then subtract the whole numbers. Remember, the fractions must have common denominators before they can be subtracted. Use the LCD.

Find: $18\frac{4}{5} - 6\frac{1}{3}$

Write the mixed numbers with like denominators. Use the LCD.	Subtract the fractions.	Subtract the whole numbers.
$18\frac{4}{5} = 18\frac{12}{15}$ $-\ 6\frac{1}{3} = 6\frac{5}{15}$	$18\frac{4}{5} = 8\frac{12}{15}$ $-\ 6\frac{1}{3} = 6\frac{5}{15}$ $\frac{7}{15}$	$18\frac{4}{5} = 18\frac{12}{15}$ $-\ 6\frac{1}{3} = 6\frac{5}{12}$ $12\frac{7}{15}$

Subtract. Write answers in lowest terms.

1. $\quad 4\frac{3}{4} = 4\frac{3}{4}$ $\qquad\qquad 1\frac{2}{3}$ $\qquad\qquad 2\frac{3}{4}$ $\qquad\qquad 4\frac{5}{9}$

$\quad -1\frac{1}{2} = 1\frac{2}{4}$ $\qquad\quad -1\frac{4}{7}$ $\qquad\quad -1\frac{1}{6}$ $\qquad\quad -1\frac{2}{6}$

$\qquad\qquad\quad 3\frac{1}{4}$

2. $\quad 11\frac{2}{3}$ $\qquad\qquad\quad 8\frac{4}{5}$ $\qquad\qquad 6\frac{2}{3}$ $\qquad\qquad 8\frac{5}{6}$

$\quad -\ 5\frac{1}{5}$ $\qquad\quad -3\frac{3}{10}$ $\qquad\quad -3\frac{1}{4}$ $\qquad\quad -3\frac{1}{2}$

3. $\quad 21\frac{2}{3}$ $\qquad\qquad\quad 5\frac{2}{3}$ $\qquad\qquad 19\frac{3}{5}$ $\qquad\qquad 33\frac{2}{4}$

$\quad -\ 3\frac{1}{8}$ $\qquad\quad -1\frac{2}{5}$ $\qquad\quad -13\frac{1}{6}$ $\qquad\quad -21\frac{1}{5}$

4. $\quad 18\frac{3}{5}$ $\qquad\qquad 15\frac{18}{20}$ $\qquad\qquad 9\frac{2}{5}$ $\qquad\qquad 84\frac{7}{12}$

$\quad -10\frac{8}{20}$ $\qquad\quad -\ 6\frac{3}{10}$ $\qquad\quad -4\frac{1}{6}$ $\qquad\quad -23\frac{1}{4}$

5. $\quad 5\frac{5}{8}$ $\qquad\qquad\quad 3\frac{4}{5}$ $\qquad\qquad 2\frac{5}{7}$ $\qquad\qquad 19\frac{3}{5}$

$\quad -2\frac{1}{4}$ $\qquad\quad -1\frac{2}{4}$ $\qquad\quad -1\frac{1}{2}$ $\qquad\quad -\ 6\frac{1}{3}$

6. $\quad 5\frac{15}{16}$ $\qquad\qquad\quad 3\frac{1}{2}$ $\qquad\qquad 7\frac{3}{8}$ $\qquad\qquad 126\frac{2}{3}$

$\quad -2\frac{3}{4}$ $\qquad\quad -2\frac{1}{7}$ $\qquad\quad -2\frac{1}{6}$ $\qquad\quad -\ 14\frac{1}{2}$

Subtracting Fractions and Mixed Numbers from Whole Numbers

Sometimes you will need to subtract a fraction or a mixed number from a whole number. To subtract from a whole number, write the whole number as a mixed number with a like denominator. Then subtract the fractions. Subtract the whole numbers.

Find: $6 - 4\frac{1}{4}$

To subtract, you need two fractions with like denominators.	Write 6 as a mixed number with 4 as the denominator.	Subtract the fractions.	Subtract the whole numbers.
$\begin{array}{r} 6 \\ -\ 4\frac{1}{4} \\ \hline \end{array}$	$6 = 5 + \frac{4}{4} = 5\frac{4}{4}$ Remember, $\frac{4}{4} = 1$	$\begin{array}{r} 6 = 5\frac{4}{4} \\ -\ 4\frac{1}{4} = 4\frac{1}{4} \\ \hline \frac{3}{4} \end{array}$	$\begin{array}{r} 6 = 5\frac{4}{4} \\ -\ 4\frac{1}{4} = 4\frac{1}{4} \\ \hline 1\frac{3}{4} \end{array}$

Write each whole number as a mixed number.

1. $12 = 11\frac{2}{2}$ $3 = 2\frac{}{3}$ $21 = 20\frac{}{4}$ $42 = 41\frac{}{2}$ $28 = 27\frac{}{6}$

2. $7 = 6\frac{}{7}$ $14 = 13\frac{}{9}$ $16 = 15\frac{}{8}$ $35 = 34\frac{}{10}$ $56 = 55\frac{}{15}$

Subtract.

3.
$\begin{array}{r} 9 = 8\frac{5}{5} \\ -\ 1\frac{2}{5} = 1\frac{2}{5} \\ \hline 7\frac{3}{5} \end{array}$
$\begin{array}{r} 7 \\ -\ 2\frac{3}{4} \\ \hline \end{array}$
$\begin{array}{r} 11 \\ -\ 8\frac{5}{6} \\ \hline \end{array}$
$\begin{array}{r} 4 \\ -\ 2\frac{1}{3} \\ \hline \end{array}$
$\begin{array}{r} 9 \\ -\ 7\frac{4}{9} \\ \hline \end{array}$

4.
$\begin{array}{r} 10 = 9\frac{11}{11} \\ -\ 2\frac{5}{11} = 2\frac{5}{11} \\ \hline 7\frac{6}{11} \end{array}$
$\begin{array}{r} 6 \\ -\ 4\frac{2}{3} \\ \hline \end{array}$
$\begin{array}{r} 19 \\ -\ 11\frac{2}{7} \\ \hline \end{array}$
$\begin{array}{r} 22 \\ -\ 16\frac{5}{6} \\ \hline \end{array}$
$\begin{array}{r} 15 \\ -\ 8\frac{7}{12} \\ \hline \end{array}$

5.
$\begin{array}{r} 8 = 7\frac{3}{3} \\ -\ \frac{1}{3} = \frac{1}{3} \\ \hline 7\frac{2}{3} \end{array}$
$\begin{array}{r} 2 \\ -\ \frac{3}{5} \\ \hline \end{array}$
$\begin{array}{r} 5 \\ -\ \frac{1}{2} \\ \hline \end{array}$
$\begin{array}{r} 3 \\ -\ \frac{1}{4} \\ \hline \end{array}$
$\begin{array}{r} 8 \\ -\ \frac{5}{6} \\ \hline \end{array}$

6.
$\begin{array}{r} 1 \\ -\ \frac{1}{2} \\ \hline \end{array}$
$\begin{array}{r} 10 \\ -\ \frac{3}{4} \\ \hline \end{array}$
$\begin{array}{r} 15 \\ -\ \frac{1}{9} \\ \hline \end{array}$
$\begin{array}{r} 1 \\ -\ \frac{1}{11} \\ \hline \end{array}$
$\begin{array}{r} 20 \\ -\ \frac{4}{15} \\ \hline \end{array}$

Subtracting Mixed Numbers with Regrouping

When subtracting mixed numbers, it may be necessary to regroup first. To regroup a mixed number for subtraction, write the whole number part as a mixed number. Add the mixed number and the fraction. Then subtract and simplify.

Find: $12\frac{1}{10} - 4\frac{3}{5}$

Write the fractions with like denominators. Compare the numerators.	$\frac{6}{10}$ is bigger than $\frac{1}{10}$. You can't subtract the fractions. To regroup, write 12 as a mixed number.	Add the mixed number and the fraction.	Now you can subtract.
$12\frac{1}{10} = 12\frac{1}{10}$ $-\ 4\frac{3}{5} = 4\frac{6}{10}$	$12 = 11\frac{10}{10}$	$12\frac{1}{10} = 11\frac{10}{10} + \frac{1}{10}$ $= 11\frac{11}{10}$ Remember, $\frac{11}{10}$ is an improper fraction.	$12\frac{1}{10} = 12\frac{1}{10} = 11\frac{11}{10}$ $-\ 4\frac{3}{5} = 4\frac{6}{10} = 4\frac{6}{10}$ $7\frac{5}{10} = 7\frac{1}{2}$

Regroup the whole number as indicated.

1. $3\frac{1}{6} = 2\frac{7}{6}$ $7\frac{1}{5} = 6\frac{}{5}$ $3\frac{2}{7} = 2\frac{}{7}$ $2\frac{5}{8} = 1\frac{}{8}$ $1\frac{1}{2} = \frac{}{2}$

Subtract.

2. $17\frac{1}{8} = 17\frac{1}{8} = 16\frac{9}{8}$ $15\frac{1}{6}$ $9\frac{1}{4}$
   $-\ 9\frac{1}{4} = 9\frac{2}{8} = 9\frac{2}{8}$ $-12\frac{1}{5}$ $-5\frac{1}{2}$
   $\qquad\qquad\qquad\qquad 7\frac{7}{8}$

3. $16\frac{1}{3}$ $15\frac{1}{12}$ $12\frac{1}{10}$
   $-\ 7\frac{1}{2}$ $-\ 9\frac{1}{4}$ $-\ 4\frac{1}{3}$

4. $10\frac{1}{15}$ $16\frac{1}{7}$ $19\frac{1}{12}$
   $-\ 4\frac{1}{5}$ $-\ 7\frac{1}{5}$ $-10\frac{1}{2}$

5. $15\frac{1}{7}$ $8\frac{1}{5}$ $12\frac{1}{5}$
   $-\ 7\frac{1}{6}$ $-3\frac{1}{2}$ $-\ 4\frac{1}{4}$

6. $15\frac{1}{2}$ $8\frac{1}{5}$ $7\frac{1}{9}$
   $-\ 7\frac{5}{9}$ $-3\frac{1}{4}$ $-4\frac{1}{3}$

Reduce to lowest terms.

1. $\frac{6}{9} =$ ___ $\frac{6}{8} =$ ___ $\frac{8}{12} =$ ___ $\frac{8}{16} =$ ___ $\frac{4}{8} =$ ___ $\frac{6}{16} =$ ___ $\frac{2}{4} =$ ___

Change to equivalent fractions.

2. $\frac{8}{12} = \frac{}{6}$ $\frac{2}{3} = \frac{}{6}$ $\frac{2}{5} = \frac{}{10}$ $\frac{4}{8} = \frac{}{2}$ $\frac{4}{6} = \frac{}{12}$ $\frac{5}{8} = \frac{}{16}$

Rename the whole number as indicated.

3. $5\frac{1}{4} = 4\frac{}{4}$ $6\frac{1}{2} = 5\frac{}{2}$ $4\frac{3}{8} = 3\frac{}{8}$ $7\frac{1}{3} = 6\frac{}{3}$ $2\frac{1}{5} = 1\frac{}{5}$ $6\frac{2}{5} = 5\frac{}{5}$

Subtract.

4.
$\frac{7}{8}$ $- \frac{1}{8}$ 　　　 $\frac{5}{6}$ $- \frac{1}{6}$ 　　　 $\frac{3}{4}$ $- \frac{1}{4}$ 　　　 $\frac{7}{8}$ $- \frac{5}{8}$ 　　　 $\frac{5}{8}$ $- \frac{3}{8}$

5.
$\frac{1}{3}$ $- \frac{1}{4}$ 　　　 $\frac{7}{8}$ $- \frac{1}{4}$ 　　　 $\frac{2}{3}$ $- \frac{1}{2}$ 　　　 $\frac{1}{3}$ $- \frac{1}{9}$ 　　　 $\frac{7}{9}$ $- \frac{1}{2}$ 　　　 $\frac{7}{12}$ $- \frac{1}{3}$

6.
$5\frac{9}{10}$ $- \frac{7}{8}$ 　　　 $6\frac{3}{8}$ $- \frac{1}{4}$ 　　　 $9\frac{6}{7}$ $- \frac{2}{3}$ 　　　 $12\frac{4}{5}$ $- \frac{1}{2}$ 　　　 $27\frac{11}{12}$ $- \frac{7}{8}$

7.
$12\frac{3}{4}$ $- 4\frac{1}{2}$ 　　　 $4\frac{1}{2}$ $- 2\frac{1}{3}$ 　　　 $6\frac{1}{2}$ $- 3\frac{1}{10}$ 　　　 $15\frac{1}{3}$ $- 4\frac{1}{8}$

8.
22 $- 12\frac{1}{2}$ 　　　 2 $- \frac{1}{3}$ 　　　 14 $- 6\frac{2}{5}$ 　　　 5 $- \frac{3}{4}$ 　　　 24 $- 13\frac{5}{12}$

9.
$8\frac{1}{6}$ $- 4\frac{3}{4}$ 　　　 $9\frac{1}{4}$ $- 2\frac{3}{4}$ 　　　 $7\frac{1}{8}$ $- 4\frac{3}{4}$ 　　　 $15\frac{1}{6}$ $- 4\frac{5}{6}$

Solve. 　　　　　　　　　　　　 Do Your Work Here

10. Dick spent $2\frac{3}{8}$ hours walking to the lake and only $1\frac{1}{8}$ hours walking back. How much faster was the trip back? _____

Problem-Solving Strategy:
Find a Pattern

The answer to a problem may be found by recognizing a pattern.
Read the problem carefully. Write the pattern and determine how
the numbers are related. Find the rule that makes the pattern.
Then solve the problem.

STEPS

1. Read the problem.

What is the next number in this number pattern?
7, 21, 63, 189, . . .

2. Determine the relationship.

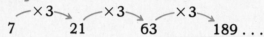

3. Write the rule.

Multiply by 3.

4. Solve the problem.

189 × 3 = 567. The next number is 567.

Solve by finding the pattern. Write the rule. Then answer the question.

1. What are the next two numbers in this
   pattern?
   5, 15, 45, 135, . . .

   Rule _____

   Answer _____

2. What are the next two numbers in this
   pattern?
   48, 44, 40, . . .

   Rule _____

   Answer _____

3. What is the next number in this pattern?
   $8\frac{2}{3}$, 10, $11\frac{1}{3}$, $12\frac{2}{3}$, . . .

   Rule _____

   Answer _____

4. What is the missing number in this pattern?
   1280, 320, _____, 20

   Rule _____

   Answer _____

Applying Your Skills

Solve. Choose the correct operation.

1. Pablo made bread and muffins. He used $\frac{3}{4}$ cup of flour for the bread and $\frac{5}{8}$ cup of flour for the muffins. How much flour did he use in all? _____

2. One bottle of cologne contains $\frac{1}{2}$ ounce. A smaller bottle contains $\frac{5}{16}$ ounce. How much more cologne does the larger bottle contain? _____

3. A share of Maxiflex stock costs $8\frac{7}{8}$ dollars. A share of Bentley stock sells for $8\frac{3}{8}$ dollars. Which stock costs less? How much less? _____ _____

4. Luis has a ring that is $\frac{3}{8}$ inch wide. Jeff has a ring that is $\frac{3}{4}$ inch wide. How much wider is Jeff's ring than Luis's ring? _____

5. Marta used $\frac{7}{8}$ cup of pecans and $\frac{1}{2}$ cup of walnuts. How many cups of nuts did she use in all? _____

6. What are the next two numbers in this pattern? 6, 12, 18, . . . _____

7. Diana bought a turkey that weighs $22\frac{1}{4}$ pounds. Her brother Joe bought one that weighs $17\frac{1}{2}$ pounds. Diana's turkey weighs how much more than Joe's? _____

8. Bill went to visit some of his friends and walked $\frac{3}{8}$ mile and $\frac{1}{4}$ mile. How far did he walk altogether? _____

9. Mae weighs $\frac{1}{2}$ pound more than Carole. Ann weighs $\frac{3}{4}$ pound more than Mae. Ann weighs how much more than Carole? _____

10. What is the next number in this pattern? $2\frac{1}{2}$, 5, $7\frac{1}{2}$, . . . _____

59

Multiplying Fractions by Fractions

To multiply fractions, multiply the numerators and multiply the denominators. Simplify the answer.

Find: $\frac{1}{7} \times \frac{4}{5}$

Multiply the numerators.

$$\frac{1}{7} \times \frac{4}{5} = \frac{1 \times 4}{} = \frac{4}{}$$

Multiply the denominators.

$$\frac{1}{7} \times \frac{4}{5} = \frac{1 \times 4}{7 \times 5} = \frac{4}{35}$$

Find: $\frac{2}{3} \times \frac{3}{8}$

Multiply the numerators.

$$\frac{2}{3} \times \frac{3}{8} = \frac{2 \times 3}{} = \frac{6}{}$$

Multiply the denominators. Simplify.

$$\frac{2}{3} \times \frac{3}{8} = \frac{2 \times 3}{3 \times 8} = \frac{6}{24} = \frac{1}{4}$$

Multiply these problems. Write answers in lowest terms.

1. $\frac{1}{3} \times \frac{1}{3} = \frac{1}{9}$ $\frac{1}{3} \times \frac{1}{4} =$ $\frac{1}{2} \times \frac{1}{2} =$ $\frac{1}{3} \times \frac{1}{2} =$

2. $\frac{1}{4} \times \frac{2}{5} = \frac{2}{20} = \frac{1}{10}$ $\frac{2}{4} \times \frac{1}{5} =$ $\frac{2}{3} \times \frac{1}{4} =$ $\frac{1}{5} \times \frac{2}{10} =$

3. $\frac{2}{9} \times \frac{2}{3} =$ $\frac{3}{5} \times \frac{2}{3} =$ $\frac{2}{3} \times \frac{3}{4} =$ $\frac{2}{5} \times \frac{5}{8} =$

4. $\frac{4}{9} \times \frac{3}{5} =$ $\frac{2}{5} \times \frac{3}{10} =$ $\frac{2}{5} \times \frac{3}{8} =$ $\frac{4}{5} \times \frac{3}{10} =$

5. $\frac{1}{3} \times \frac{2}{3} =$ $\frac{4}{5} \times \frac{1}{4} =$ $\frac{1}{5} \times \frac{3}{10} =$ $\frac{1}{6} \times \frac{1}{3} =$

6. $\frac{3}{4} \times \frac{1}{4} =$ $\frac{7}{10} \times \frac{1}{5} =$ $\frac{3}{8} \times \frac{1}{6} =$ $\frac{2}{3} \times \frac{3}{7} =$

7. $\frac{2}{5} \times \frac{1}{5} =$ $\frac{1}{3} \times \frac{3}{5} =$ $\frac{1}{7} \times \frac{2}{7} =$ $\frac{2}{9} \times \frac{1}{2} =$

8. $\frac{3}{4} \times \frac{4}{5} =$ $\frac{3}{10} \times \frac{1}{3} =$ $\frac{1}{2} \times \frac{2}{7} =$ $\frac{2}{5} \times \frac{3}{7} =$

Solve.

Do Your Work Here

9. For a picnic, we made ice cream. We used an 8-quart freezer, and it was three fourths full when the ice cream was frozen. How many quarts of ice cream did we have? _____

10. Mrs. Denman bought a loaf of bread which weighed three fourths of a pound. If her family ate half of the loaf, how much bread was left? _____

11. Marsha lived $\frac{9}{16}$ mile from town. Liz lived only one half as far from town. How far did Liz live from town? _____

Using Cancellation

Instead of simplifying fractions after they have been multiplied, it may be possible to use **cancellation** before multiplying. To cancel, divide the numerator and the denominator by the same number, or common factor. Then multiply, using the new numerator and denominator.

Find: $\frac{3}{10} \times \frac{1}{9}$

Find a number which will divide into both the 3 and the 9.	Cancel by dividing both the 3 and the 9 by 3.	Multiply the new numerators and denominators.
$\frac{3}{10} \times \frac{1}{9}$	$\frac{\overset{1}{\cancel{3}}}{10} \times \frac{1}{\underset{3}{\cancel{9}}}$	$\frac{1 \times 1}{10 \times 3} = \frac{1}{30}$

Work the following problems using cancellation.

1. $\frac{1}{3} \times \frac{6}{7} =$ $\frac{3}{5} \times \frac{1}{9} =$ $\frac{2}{5} \times \frac{3}{4} =$ $\frac{4}{5} \times \frac{15}{16} =$ $\frac{1}{6} \times \frac{3}{7} =$

$\frac{1}{\underset{1}{\cancel{3}}} \times \frac{\overset{2}{\cancel{6}}}{7} = \frac{1 \times 2}{1 \times 7} = \frac{2}{7}$

2. $\frac{2}{5} \times \frac{5}{8} =$ $\frac{1}{4} \times \frac{2}{5} =$ $\frac{1}{3} \times \frac{3}{7} =$ $\frac{3}{5} \times \frac{10}{16} =$ $\frac{2}{3} \times \frac{3}{4} =$

3. $\frac{5}{9} \times \frac{3}{10} =$ $\frac{7}{8} \times \frac{2}{7} =$ $\frac{5}{12} \times \frac{4}{15} =$ $\frac{2}{7} \times \frac{7}{9} =$ $\frac{3}{8} \times \frac{4}{9} =$

$\frac{\overset{1}{\cancel{5}}}{\underset{3}{\cancel{9}}} \times \frac{\overset{1}{\cancel{3}}}{\underset{2}{\cancel{10}}} = \frac{1 \times 1}{3 \times 2} = \frac{1}{6}$

4. $\frac{3}{4} \times \frac{4}{5} =$ $\frac{2}{3} \times \frac{6}{7} =$ $\frac{6}{7} \times \frac{5}{12} =$ $\frac{5}{11} \times \frac{3}{10} =$ $\frac{4}{7} \times \frac{7}{9} =$

5. $\frac{4}{9} \times \frac{3}{4} =$ $\frac{5}{6} \times \frac{3}{10} =$ $\frac{5}{8} \times \frac{4}{9} =$ $\frac{3}{5} \times \frac{5}{9} =$ $\frac{7}{8} \times \frac{4}{7} =$

6. $\frac{7}{12} \times \frac{4}{7} =$ $\frac{1}{8} \times \frac{8}{9} =$ $\frac{3}{4} \times \frac{1}{6} =$ $\frac{5}{7} \times \frac{7}{15} =$ $\frac{3}{16} \times \frac{8}{9} =$

Solve the following word problems.

7. Terry lives $\frac{6}{8}$ mile from town. Patricia lives halfway between Terry and town. How far from town does Patricia live?   _____

8. Ms. Miller lives $\frac{9}{16}$ mile from her office. Mr. Brown lives only one third as far from his office. How far is it to Mr. Brown's office?   _____

61

Multiplying Fractions and Whole Numbers

To multiply a whole number and a fraction, first write the whole number as an improper fraction. Cancel if possible. Multiply the new numerators and denominators. Simplify the answer.

Find: $\frac{4}{9} \times 3$

Write the whole number as an improper fraction.	Cancel.	Multiply the new numerators and denominators.	Simplify.
$\frac{4}{9} \times 3 = \frac{4}{9} \times \frac{3}{1}$	$\frac{4}{\underset{3}{9}} \times \frac{\overset{1}{3}}{1}$	$\frac{4 \times 1}{3 \times 1} = \frac{4}{3}$	$\frac{4}{3} = 1\frac{1}{3}$

Solve the following problems. Simplify where possible.

1. $4 \times \frac{1}{3} = \frac{4}{1} \times \frac{1}{3} = \frac{4}{3} = 1\frac{1}{3}$ $5 \times \frac{1}{4} =$ $6 \times \frac{1}{5} =$

2. $4 \times \frac{2}{3} =$ $5 \times \frac{2}{4} =$ $6 \times \frac{2}{5} =$

3. $7 \times \frac{1}{2} =$ $9 \times \frac{3}{4} =$ $6 \times \frac{4}{5} =$

4. $21 \times \frac{1}{3} = \frac{\overset{7}{21}}{1} \times \frac{1}{\underset{1}{3}} = \frac{7}{1} = 7$ $15 \times \frac{2}{3} =$ $18 \times \frac{5}{6} =$

5. $20 \times \frac{2}{5} =$ $18 \times \frac{2}{9} =$ $25 \times \frac{2}{5} =$

6. $\frac{1}{4} \times 4 = \frac{1}{\underset{1}{4}} \times \frac{\overset{1}{4}}{1} = \frac{1}{1} = 1$ $\frac{1}{3} \times 15 =$ $\frac{2}{5} \times 125 =$

7. $24 \times \frac{5}{8} =$ $75 \times \frac{2}{5} =$ $50 \times \frac{3}{10} =$

8. $\frac{3}{8} \times 24 =$ $\frac{2}{5} \times 45 =$ $\frac{2}{3} \times 30 =$

9. $21 \times \frac{2}{7} =$ $4 \times \frac{5}{7} =$ $40 \times \frac{3}{8} =$

10. $35 \times \frac{3}{5} =$ $15 \times \frac{4}{5} =$ $17 \times \frac{2}{3} =$

11. $12 \times \frac{5}{6} =$ $18 \times \frac{4}{9} =$ $20 \times \frac{3}{5} =$

Solve.

Do Your Work Here

12. Mr. Green wished to make ribbons for each of his five girls. Each ribbon would be $\frac{3}{4}$ yard long. How many yards did he need in all?

_____

Multiplying Mixed Numbers and Whole Numbers

To multiply a mixed number and a whole number, write the mixed number and the whole number as improper fractions. Use cancellation if possible. Multiply the new numerators and denominators. Simplify the answer.

Find: $1\frac{3}{5} \times 15$

Write the whole number and the mixed number as improper fractions.	Cancel.	Multiply the new numerators and denominators. Simplify.
$1\frac{3}{5} \times 15 = \frac{8}{5} \times \frac{15}{1}$	$\frac{8}{5} \times \frac{\overset{3}{\cancel{15}}}{\underset{1}{1}}$	$\frac{8 \times 3}{1 \times 1} = \frac{24}{1} = 24$

Multiply. Use cancellation if possible. Simplify.

1. $1\frac{1}{2} \times 4 =$ $4\frac{3}{4} \times 3 =$ $2\frac{3}{5} \times 10 =$

$\frac{3}{\underset{1}{\cancel{2}}} \times \frac{\overset{2}{\cancel{4}}}{1} = \frac{3 \times 2}{1 \times 1} = \frac{6}{1} = 6$

2. $2\frac{1}{3} \times 9 =$ $1\frac{4}{5} \times 5 =$ $3\frac{2}{3} \times 12 =$

3. $3\frac{14}{15} \times 2 =$ $9\frac{1}{10} \times 11 =$ $4\frac{5}{6} \times 9 =$

$\frac{59}{15} \times \frac{2}{1} = \frac{118}{15} = 7\frac{13}{15}$

4. $12\frac{1}{10} \times 20 =$ $7\frac{5}{6} \times 12 =$ $8\frac{1}{5} \times 15 =$

5. $3 \times 4\frac{3}{4} =$ $6 \times 4\frac{5}{21} =$ $4 \times 1\frac{7}{8} =$

$\frac{3}{1} \times \frac{19}{4} = \frac{57}{4} = 14\frac{1}{4}$

6. $1 \times 4\frac{9}{10} =$ $7 \times 2\frac{3}{10} =$ $8 \times 6\frac{1}{2} =$

7. $14 \times 1\frac{3}{8} =$ $5\frac{1}{6} \times 2 =$ $12 \times 4\frac{1}{4} =$

8. $10 \times 2\frac{3}{4} =$ $1\frac{5}{6} \times 3 =$ $6 \times 5\frac{7}{12} =$

Multiplying Mixed Numbers and Fractions

To multiply a mixed number and a fraction, write the mixed number as an improper fraction. Use cancellation if possible. Multiply the new numerators and denominators. Simplify the answer.

Find: $\frac{2}{3} \times 5\frac{1}{4}$

Write $5\frac{1}{4}$ as an improper fraction. Multiply the denominator 4 by the whole number: $4 \times 5 = 20$. Add the numerator: $20 + 1 = 21$. Place this new numerator over the old denominator: $5\frac{1}{4} = \frac{21}{4}$	Write the mixed number as an improper fraction. $\frac{2}{3} \times 5\frac{1}{4} = \frac{2}{3} \times \frac{21}{4}$	Cancel. $\frac{\overset{1}{\cancel{2}}}{\underset{1}{\cancel{3}}} \times \frac{\overset{7}{\cancel{21}}}{\underset{2}{\cancel{4}}} =$	Multiply the new numerators and denominators. Simplify. $\frac{1 \times 7}{1 \times 2} = \frac{7}{2} = 3\frac{1}{2}$

Multiply. Use cancellation if possible. Simplify.

1. $\frac{1}{2} \times 3\frac{1}{2} =$
$\frac{1}{2} \times \frac{7}{2} = \frac{1 \times 7}{2 \times 2} = \frac{7}{4} = 1\frac{3}{4}$

$\frac{3}{5} \times 3\frac{1}{3} =$

$\frac{4}{5} \times 1\frac{1}{2} =$

2. $\frac{2}{3} \times 5\frac{1}{4} =$ $\frac{2}{9} \times 4\frac{1}{2} =$ $\frac{3}{8} \times 2\frac{1}{9} =$

3. $\frac{1}{2} \times 1\frac{3}{5} =$ $\frac{1}{2} \times 3\frac{1}{2} =$ $\frac{5}{6} \times 3\frac{2}{5} =$

4. $\frac{12}{23} \times 5\frac{3}{4} =$ $\frac{9}{16} \times 2\frac{2}{3} =$ $\frac{7}{12} \times 4\frac{3}{7} =$

5. $3\frac{1}{2} \times \frac{3}{8} =$ $1\frac{7}{8} \times \frac{4}{15} =$ $2\frac{1}{3} \times \frac{4}{7} =$

6. $7\frac{2}{3} \times \frac{1}{2} =$ $4\frac{2}{3} \times \frac{3}{7} =$ $5\frac{1}{2} \times \frac{2}{11} =$

7. $2\frac{1}{2} \times \frac{1}{3} =$ $9\frac{1}{2} \times \frac{1}{8} =$ $6\frac{3}{4} \times \frac{2}{9} =$

8. $2\frac{1}{3} \times \frac{6}{7} =$ $3\frac{3}{4} \times \frac{7}{12} =$ $1\frac{5}{6} \times \frac{3}{4} =$

9. $\frac{2}{9} \times 3\frac{3}{8} =$ $4\frac{1}{4} \times \frac{4}{17} =$ $3\frac{5}{8} \times \frac{5}{6} =$

Multiplying Mixed Numbers
by Mixed Numbers

To multiply a mixed number by a mixed number, write both mixed numbers as improper fractions. Use cancellation if possible. Multiply the new numerators and denominators. Simplify the answer.

Find: $1\frac{2}{3} \times 4\frac{1}{2}$

Write the mixed numbers as improper fractions.	Cancel.	Multiply the new numerators and denominators. Simplify.
$1\frac{2}{3} \times 4\frac{1}{2} = \frac{5}{3} \times \frac{9}{2}$	$\frac{5}{3} \times \frac{\overset{3}{\cancel{9}}}{2}$ $\underset{1}{\cancel{3}}$	$\frac{5 \times 3}{1 \times 2} = \frac{15}{2} = 7\frac{1}{2}$

Multiply. Use cancellation if possible. Simplify.

1. $2\frac{3}{8} \times 2\frac{1}{3} =$ 　　　　$1\frac{1}{3} \times 2\frac{1}{2} =$ 　　　　$3\frac{1}{2} \times 2\frac{1}{3} =$

$\frac{19}{8} \times \frac{7}{3} = \frac{19 \times 7}{8 \times 3} = \frac{133}{24} = 5\frac{13}{24}$

2. $2\frac{2}{5} \times 4\frac{1}{2} =$ 　　　　$2\frac{1}{5} \times 2\frac{1}{4} =$ 　　　　$1\frac{3}{4} \times 4\frac{1}{2} =$

3. $1\frac{1}{14} \times 1\frac{3}{4} =$ 　　　　$3\frac{3}{4} \times 2\frac{2}{3} =$ 　　　　$2\frac{1}{5} \times 3\frac{1}{8} =$

4. $3\frac{4}{7} \times 2\frac{4}{5} =$ 　　　　$2\frac{3}{4} \times 5\frac{1}{4} =$ 　　　　$1\frac{1}{6} \times 1\frac{1}{8} =$

5. $8\frac{3}{9} \times 1\frac{2}{25} =$ 　　　　$6\frac{1}{2} \times 1\frac{1}{3} =$ 　　　　$1\frac{4}{5} \times 3\frac{3}{4} =$

6. $2\frac{1}{3} \times 5\frac{1}{5} =$ 　　　　$2\frac{1}{12} \times 3\frac{6}{25} =$ 　　　　$3\frac{1}{6} \times 3\frac{2}{5} =$

7. $1\frac{4}{5} \times 1\frac{2}{9} =$ 　　　　$2\frac{5}{6} \times 3\frac{3}{4} =$ 　　　　$4\frac{5}{9} \times 2\frac{1}{5} =$

8. $2\frac{2}{7} \times 1\frac{2}{5} =$ 　　　　$4\frac{3}{5} \times 2\frac{1}{3} =$ 　　　　$6\frac{1}{2} \times 3\frac{1}{3} =$

9. $4\frac{4}{5} \times 3\frac{1}{8} =$ 　　　　$2\frac{4}{10} \times 3\frac{1}{3} =$ 　　　　$4\frac{5}{7} \times 1\frac{7}{8} =$

Checking Up

Simplify these improper fractions.

1. $\frac{9}{2}$ = _____ $\frac{17}{5}$ = _____ $\frac{10}{3}$ = _____ $\frac{20}{6}$ = _____ $\frac{8}{3}$ = _____ $\frac{24}{7}$ = _____

2. $\frac{10}{4}$ = _____ $\frac{15}{10}$ = _____ $\frac{16}{3}$ = _____ $\frac{12}{5}$ = _____ $\frac{21}{8}$ = _____ $\frac{19}{4}$ = _____

Change to an improper fraction.

3. $4\frac{1}{2}$ = _____ 5 = _____ $6\frac{1}{3}$ = _____ $7\frac{1}{4}$ = _____ 3 = _____ $4\frac{1}{8}$ = _____

4. $5\frac{2}{5}$ = _____ $7\frac{3}{4}$ = _____ 9 = _____ $6\frac{3}{5}$ = _____ $15\frac{3}{10}$ = _____ $12\frac{5}{8}$ = _____

Multiply. Cancel when possible.

5. $\frac{4}{7} \times \frac{3}{4}$ = $\frac{9}{10} \times \frac{2}{3}$ = $\frac{5}{6} \times \frac{2}{5}$ = $\frac{3}{4} \times \frac{5}{6}$ = $\frac{5}{6} \times \frac{5}{8}$ =

6. $\frac{1}{2} \times \frac{2}{3}$ = $\frac{1}{2} \times \frac{4}{5}$ = $\frac{3}{8} \times \frac{4}{9}$ = $\frac{2}{5} \times \frac{15}{32}$ = $\frac{3}{16} \times \frac{8}{9}$ =

7. $18 \times \frac{1}{2}$ = $15 \times \frac{1}{3}$ = $12 \times \frac{1}{4}$ = $32 \times \frac{1}{8}$ = $45 \times \frac{1}{5}$ =

8. $\frac{2}{3} \times 24$ = $\frac{3}{4} \times 24$ = $\frac{2}{5} \times 15$ = $\frac{3}{8} \times 40$ = $\frac{7}{10} \times 50$ =

9. $\frac{1}{3} \times 6\frac{1}{2}$ = $2\frac{1}{2} \times \frac{2}{5}$ = $8 \times 8\frac{1}{4}$ = $4\frac{3}{4} \times 8$ = $\frac{1}{2} \times 2\frac{5}{9}$ =

10. $12\frac{1}{2} \times 2\frac{2}{5}$ = $\frac{1}{4} \times 3\frac{1}{3}$ = $3\frac{1}{3} \times 7\frac{1}{2}$ = $6 \times 8\frac{3}{4}$ = $8\frac{3}{4} \times 9\frac{2}{7}$ =

Solve.

11. Anna Polansky bought $\frac{3}{4}$ yard of dress material at $4.00 per yard. How much did she pay? _____

12. Bert Williams made kitchen towels $\frac{2}{3}$ yard in length. How much material did he need for a dozen kitchen towels? _____

13. Joe bought $3\frac{1}{2}$ quarts of strawberries at the store for $2.00 a quart. How much did he have to pay for them? _____

14. The baker used 12 one-quarter-pound bars of chocolate for a recipe which called for 3 pounds. Did the baker use the right amount? _____

66

Dividing Fractions by Whole Numbers

Two thirds of the rectangle at the right is shaded. If this shaded part is divided into two equal parts, what portion of the whole rectangle will **one** part be?

$$\tfrac{2}{3} \div 2 = \text{_____ third}$$

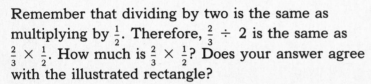

Remember that dividing by two is the same as multiplying by $\frac{1}{2}$. Therefore, $\frac{2}{3} \div 2$ is the same as $\frac{2}{3} \times \frac{1}{2}$. How much is $\frac{2}{3} \times \frac{1}{2}$? Does your answer agree with the illustrated rectangle?

The rectangle at the right is divided into 8 parts. Six of these parts are shaded.

If the shaded area is divided by 2, what portion of the whole rectangle will one of these two parts be?

$$\tfrac{6}{8} \div 2 = \text{_____ eighths}$$

Again, remembering that dividing by 2 is the same as multiplying by $\frac{1}{2}$, change $\frac{6}{8} \div 2$ to read $\frac{6}{8} \times \frac{1}{2}$. How much is $\frac{6}{8} \times \frac{1}{2}$? Does this agree with the illustration at the right?

In dividing a fraction by a whole number, if the whole number will divide evenly into the numerator, you can use that procedure. In the first illustration, 2 is contained in the numerator 2 exactly once, giving an answer of $\frac{1}{3}$. In the second illustration, 2 goes into the numerator 6 exactly three times, giving an answer of $\frac{3}{8}$. There is one rule that fits all occasions:

To divide with fractions, invert the divisor and multiply. Invert means to turn over. Thus, in the above illustrations, 2 inverted becomes $\frac{1}{2}$, since 2 can be written in fraction form as $\frac{2}{1}$.

Complete these sentences.

1. 2 inverted is ___.   3 inverted is ___.   5 inverted is ___.   10 inverted is ___.

2. $\frac{1}{2}$ inverted is ___. $\frac{1}{3}$ inverted is ___. $\frac{1}{4}$ inverted is ___. $\frac{1}{5}$ inverted is ___.

3. $\frac{2}{3}$ inverted is ___. $\frac{3}{4}$ inverted is ___. $\frac{5}{8}$ inverted is ___. $\frac{7}{8}$ inverted is ___.

Divide. Remember to invert the divisor and multiply. Cancel if possible.

4. $\frac{8}{9} \div 2 =$ $\frac{2}{3} \div 2 =$ $\frac{3}{8} \div 3 =$ $\frac{5}{8} \div 5 =$ $\frac{2}{5} \div 4 =$

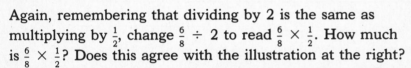

$$\overset{4}{\cancel{\tfrac{8}{9}}} \times \underset{1}{\tfrac{1}{\cancel{2}}} = \tfrac{4 \times 1}{9 \times 1} = \tfrac{4}{9}$$

5. $\frac{4}{5} \div 2 =$ $\frac{5}{8} \div 10 =$ $\frac{5}{6} \div 12 =$ $\frac{7}{8} \div 3 =$ $\frac{5}{9} \div 6 =$

6. $\frac{3}{7} \div 4 =$ $\frac{5}{18} \div 3 =$ $\frac{5}{11} \div 2 =$ $\frac{7}{12} \div 3 =$ $\frac{9}{16} \div 4 =$

Dividing Fractions by Fractions

To divide a fraction by a fraction, follow the division rule: Invert the divisor and multiply. Remember, to invert means to turn over. Only the second fraction is inverted. Cancel if possible and simplify.

Find: $\frac{2}{3} \div \frac{5}{12}$

Invert the divisor and multiply.	Cancel.	Multiply the new numerators and denominators. Simplify.
$\frac{2}{3} \div \frac{5}{12} = \frac{2}{3} \times \frac{12}{5}$	$\frac{2}{\cancel{3}} \times \frac{\cancel{12}^{4}}{5}$	$\frac{2 \times 4}{1 \times 5} = \frac{8}{5} = 1\frac{3}{5}$

Work these examples. Cancel when possible.

1. $\frac{1}{4} \div \frac{1}{8} =$ $\frac{1}{3} \div \frac{1}{9} =$ $\frac{1}{8} \div \frac{1}{16} =$ $\frac{1}{6} \div \frac{1}{12} =$ $\frac{1}{5} \div \frac{1}{10} =$

$\frac{1}{\cancel{4}} \times \frac{\cancel{8}^{2}}{1} = \frac{1 \times 2}{1 \times 1} = \frac{2}{1} = 2$

2. $\frac{3}{4} \div \frac{5}{8} =$ $\frac{4}{5} \div \frac{1}{10} =$ $\frac{3}{8} \div \frac{3}{4} =$ $\frac{3}{7} \div \frac{3}{5} =$ $\frac{2}{3} \div \frac{4}{5} =$

3. $\frac{2}{9} \div \frac{3}{4} =$ $\frac{5}{12} \div \frac{3}{4} =$ $\frac{9}{16} \div \frac{3}{8} =$ $\frac{5}{6} \div \frac{5}{8} =$ $\frac{7}{8} \div \frac{5}{12} =$

4. $\frac{5}{16} \div \frac{5}{32} =$ $\frac{5}{64} \div \frac{5}{16} =$ $\frac{10}{64} \div \frac{1}{4} =$ $\frac{3}{16} \div \frac{9}{32} =$ $\frac{7}{12} \div \frac{3}{4} =$

Solve the following word problems.

Do Your Work Here

5. A chocolate bar weighing $\frac{3}{4}$ pound is divided into squares, each square weighing $\frac{1}{32}$ pound. How many squares are there in the bar? _____

6. The cafe owner used a $\frac{3}{4}$ pound can of pepper to fill the pepper shakers. Each shaker would hold $\frac{1}{64}$ pound of pepper. In this way, how many shakers could he fill from the $\frac{3}{4}$ pound can? _____

7. A small box of candy bars weighs $\frac{15}{16}$ pound. Each bar weighs $\frac{3}{32}$ pound. How many bars are there in this box? _____

Dividing Whole Numbers by Fractions

To divide a whole number by a fraction, write the whole number as an improper fraction. Invert the divisor and multiply. Cancel if possible. Simplify.

Find: $6 \div \frac{2}{5}$

Write the whole number as an improper fraction.	Invert the divisor and multiply.	Cancel.	Multiply. Simplify.
$6 \div \frac{2}{5} = \frac{6}{1} \div \frac{2}{5}$	$\frac{6}{1} \times \frac{5}{2}$	$\frac{\overset{3}{\cancel{6}}}{1} \times \frac{5}{\underset{1}{\cancel{2}}}$	$\frac{3 \times 5}{1 \times 1} = \frac{15}{1} = 15$

Divide. Cancel when possible. Simplify.

1. $5 \div \frac{1}{5} =$ $\qquad$ $4 \div \frac{1}{4} =$ $\qquad$ $9 \div \frac{1}{3} =$ $\qquad$ $8 \div \frac{1}{2} =$

$\frac{5}{1} \times \frac{5}{1} = \frac{5 \times 5}{1 \times 1} = \frac{25}{1} = 25$

2. $5 \div \frac{1}{3} =$ $\qquad$ $12 \div \frac{3}{4} =$ $\qquad$ $9 \div \frac{3}{5} =$ $\qquad$ $16 \div \frac{2}{3} =$

3. $30 \div \frac{5}{6} =$ $\qquad$ $18 \div \frac{3}{4} =$ $\qquad$ $21 \div \frac{7}{9} =$ $\qquad$ $40 \div \frac{5}{8} =$

$\frac{\overset{6}{\cancel{30}}}{1} \times \frac{6}{\underset{1}{\cancel{5}}} = \frac{6 \times 6}{1 \times 1} = \frac{36}{1} = 36$

4. $17 \div \frac{2}{3} =$ $\qquad$ $21 \div \frac{6}{7} =$ $\qquad$ $15 \div \frac{3}{7} =$ $\qquad$ $24 \div \frac{4}{7} =$

5. $25 \div \frac{5}{7} =$ $\qquad$ $50 \div \frac{5}{8} =$ $\qquad$ $45 \div \frac{9}{10} =$ $\qquad$ $36 \div \frac{3}{4} =$

6. $35 \div \frac{7}{9} =$ $\qquad$ $40 \div \frac{8}{9} =$ $\qquad$ $50 \div \frac{5}{12} =$ $\qquad$ $28 \div \frac{4}{7} =$

7. $72 \div \frac{8}{9} =$ $\qquad$ $63 \div \frac{9}{10} =$ $\qquad$ $96 \div \frac{24}{25} =$ $\qquad$ $100 \div \frac{25}{32} =$

Solve these word problems.

░░░░ **Do Your Work Here** ░░░░

8. The grocer bought a 12-pound carton of pepper. The pepper was in $\frac{1}{8}$-pound cans. How many cans were in the carton? _____

9. For a gardening project, a 4-acre vacant plot of ground was divided into $\frac{1}{4}$-acre plots. How many plots were made? _____

Dividing with Mixed Numbers and Whole Numbers

To divide with mixed numbers and whole numbers, write the mixed numbers and whole numbers as improper fractions. Invert the divisor and multiply. Cancel when possible. Simplify if needed.

Find: $2\frac{5}{8} \div 3$

Write the mixed number and whole number as improper fractions.	Invert the divisor and multiply.	Cancel.	Multiply.
$2\frac{5}{8} \div 3 = \frac{21}{8} \div \frac{3}{1}$	$\frac{21}{8} \times \frac{1}{3}$	$\frac{\overset{7}{21}}{8} \times \frac{1}{\underset{1}{3}}$	$\frac{7 \times 1}{8 \times 1} = \frac{7}{8}$

Divide. Cancel if possible. Simplify.

1. $3\frac{1}{2} \div 7 =$ $\qquad$ $1\frac{2}{3} \div 5 =$ $\qquad$ $2\frac{3}{4} \div 11 =$

$\frac{7}{2} \div \frac{7}{1} = \frac{\overset{1}{\cancel{7}}}{2} \times \frac{1}{\underset{1}{\cancel{7}}} = \frac{1 \times 1}{2 \times 1} = \frac{1}{2}$

2. $4\frac{1}{2} \div 3 =$ $\qquad$ $4\frac{1}{8} \div 3 =$ $\qquad$ $9\frac{1}{3} \div 7 =$

$\frac{9}{2} \div \frac{3}{1} = \frac{\overset{3}{\cancel{9}}}{2} \times \frac{1}{\underset{1}{\cancel{3}}} = \frac{3 \times 1}{2 \times 1} = \frac{3}{2} = 1\frac{1}{2}$

3. $3\frac{1}{7} \div 11 =$ $\qquad$ $7\frac{1}{2} \div 3 =$ $\qquad$ $1\frac{7}{8} \div 5 =$

4. $5 \div 3\frac{1}{3} =$ $\qquad$ $13 \div 5\frac{4}{7} =$ $\qquad$ $8 \div 5\frac{1}{3} =$

$\frac{5}{1} \div \frac{10}{3} = \frac{\overset{1}{\cancel{5}}}{1} \times \frac{3}{\underset{2}{\cancel{10}}} = \frac{1 \times 3}{1 \times 2} = \frac{3}{2} = 1\frac{1}{2}$

5. $2 \div 6\frac{2}{3} =$ $\qquad$ $4 \div 2\frac{2}{3} =$ $\qquad$ $5 \div 4\frac{1}{6} =$

6. $3\frac{3}{4} \div 12 =$ $\qquad$ $2 \div 1\frac{1}{3} =$ $\qquad$ $2\frac{1}{2} \div 5 =$

7. $14 \div 3\frac{1}{9} =$ $\qquad$ $4\frac{2}{5} \div 2 =$ $\qquad$ $2 \div 7\frac{1}{3} =$

8. $11 \div 8\frac{1}{4} =$ $\qquad$ $2\frac{1}{4} \div 6 =$ $\qquad$ $4 \div 3\frac{1}{5} =$

9. $3\frac{1}{2} \div 5 =$ $\qquad$ $5 \div 2\frac{1}{2} =$ $\qquad$ $6\frac{2}{3} \div 10 =$

10. $11 \div 4\frac{1}{8} =$ $\qquad$ $5\frac{1}{3} \div 24 =$ $\qquad$ $39 \div 8\frac{2}{3} =$

70

Dividing Mixed Numbers by Fractions

To divide a mixed number by a fraction, write the mixed number as an improper fraction. Invert the divisor and multiply. Cancel if possible. Simplify the answer.

Find: $6\frac{5}{6} \div \frac{5}{6}$

Write the mixed number as an improper fraction.	Invert the divisor and multiply.	Cancel.	Multiply. Simplify.
$6\frac{5}{6} \div \frac{5}{6} = \frac{41}{6} \div \frac{5}{6}$	$\frac{41}{6} \times \frac{6}{5}$	$\frac{41}{\cancel{6}} \times \frac{\cancel{6}}{5}$	$\frac{41 \times 1}{1 \times 5} = \frac{41}{5} = 8\frac{1}{5}$

Divide. Cancel if possible. Simplify.

1. $2\frac{1}{3} \div \frac{1}{6} =$ $1\frac{1}{3} \div \frac{2}{3} =$ $3\frac{1}{2} \div \frac{1}{4} =$

$\frac{7}{3} \div \frac{1}{6} = \frac{7}{\cancel{3}} \times \frac{\cancel{6}}{1} = \frac{7 \times 2}{1 \times 1} = \frac{14}{1} = 14$

2. $1\frac{7}{12} \div \frac{3}{4} =$ $3\frac{1}{10} \div \frac{2}{5} =$ $2\frac{3}{4} \div \frac{3}{8} =$

$\frac{19}{12} \div \frac{3}{4} = \frac{19}{\cancel{12}} \times \frac{\cancel{4}}{3} = \frac{19 \times 1}{3 \times 3} = \frac{19}{9} = 2\frac{1}{9}$

3. $5\frac{1}{2} \div \frac{1}{2} =$ $3\frac{3}{4} \div \frac{1}{2} =$ $4\frac{1}{3} \div \frac{5}{6} =$

4. $1\frac{1}{2} \div \frac{3}{4} =$ $1\frac{1}{5} \div \frac{4}{5} =$ $3\frac{1}{4} \div \frac{5}{12} =$

5. $2\frac{1}{5} \div \frac{4}{5} =$ $2\frac{2}{9} \div \frac{4}{5} =$ $1\frac{5}{6} \div \frac{2}{3} =$

6. $1\frac{4}{5} \div \frac{2}{3} =$ $1\frac{4}{7} \div \frac{1}{7} =$ $4\frac{3}{5} \div \frac{2}{5} =$

7. $5\frac{1}{4} \div \frac{1}{3} =$ $3\frac{2}{3} \div \frac{11}{12} =$ $2\frac{1}{8} \div \frac{1}{4} =$

8. $3\frac{3}{4} \div \frac{5}{8} =$ $5\frac{2}{5} \div \frac{3}{5} =$ $2\frac{4}{9} \div \frac{2}{3} =$

9. $2\frac{3}{4} \div \frac{8}{9} =$ $4\frac{7}{8} \div \frac{7}{8} =$ $5\frac{1}{7} \div \frac{5}{14} =$

Dividing Mixed Numbers
by Mixed Numbers

To divide a mixed number by a mixed number, write the mixed numbers as improper fractions. Invert the divisor and multiply. Cancel if possible. Simplify if needed.

Find: $2\frac{1}{4} \div 3\frac{3}{8}$

Write the mixed numbers as improper fractions.	Invert the divisor and multiply.	Cancel.	Multiply.
$2\frac{1}{4} \div 3\frac{3}{8} = \frac{9}{4} \div \frac{27}{8}$	$\frac{9}{4} \times \frac{8}{27}$	$\overset{1}{\underset{1}{\cancel{\frac{9}{4}}}} \times \overset{2}{\underset{3}{\cancel{\frac{8}{27}}}}$	$\frac{1 \times 2}{1 \times 3} = \frac{2}{3}$

Divide. Cancel if possible. Simplify.

1. $4\frac{1}{4} \div 8\frac{1}{2} =$ $\qquad$ $3\frac{1}{3} \div 1\frac{20}{21} =$ $\qquad$ $2\frac{3}{4} \div 4\frac{1}{2} =$

$\frac{17}{4} \div \frac{17}{2} = \overset{1}{\underset{2}{\cancel{\frac{17}{4}}}} \times \overset{1}{\underset{1}{\cancel{\frac{2}{17}}}} = \frac{1 \times 1}{2 \times 1} = \frac{1}{2}$

2. $5\frac{1}{2} \div 1\frac{1}{2} =$ $\qquad$ $7\frac{1}{2} \div 4\frac{3}{5} =$ $\qquad$ $3\frac{1}{2} \div 2\frac{1}{2} =$

$\frac{11}{2} \div \frac{3}{2} = \frac{11}{\underset{1}{\cancel{2}}} \times \frac{\cancel{2}}{3} = \frac{11 \times 1}{1 \times 3} = \frac{11}{3} = 3\frac{2}{3}$

3. $8\frac{1}{4} \div 2\frac{1}{2} =$ $\qquad$ $2\frac{1}{3} \div 3\frac{1}{2} =$ $\qquad$ $6\frac{1}{3} \div 1\frac{5}{6} =$

4. $6\frac{2}{3} \div 2\frac{1}{5} =$ $\qquad$ $8\frac{2}{3} \div 1\frac{1}{3} =$ $\qquad$ $5\frac{1}{3} \div 1\frac{1}{3} =$

5. $8\frac{1}{2} \div 4\frac{1}{4} =$ $\qquad$ $6\frac{4}{5} \div 1\frac{1}{5} =$ $\qquad$ $4\frac{3}{4} \div 1\frac{1}{8} =$

6. $4\frac{1}{2} \div 1\frac{1}{4} =$ $\qquad$ $9\frac{7}{9} \div 1\frac{5}{6} =$ $\qquad$ $3\frac{3}{8} \div 1\frac{1}{4} =$

7. $2\frac{1}{2} \div 1\frac{1}{3} =$ $\qquad$ $1\frac{3}{8} \div 3\frac{2}{3} =$ $\qquad$ $4\frac{1}{5} \div 1\frac{2}{5} =$

8. $6\frac{2}{3} \div 2\frac{1}{4} =$ $\qquad$ $4\frac{2}{9} \div 1\frac{7}{12} =$ $\qquad$ $2\frac{1}{7} \div 4\frac{2}{7} =$

9. $4\frac{3}{4} \div 1\frac{1}{8} =$ $\qquad$ $4\frac{2}{3} \div 3\frac{1}{2} =$ $\qquad$ $2\frac{1}{4} \div 3\frac{3}{8} =$

Checking Up

Change to improper fractions.

1. $4\frac{1}{3} =$ $5 =$ $6\frac{2}{5} =$ $8 =$ $15\frac{2}{3} =$ $25\frac{3}{4} =$ $17 =$ $100\frac{1}{2} =$

Rename these improper fractions as mixed numbers.

2. $\frac{17}{3} =$ $\frac{19}{4} =$ $\frac{31}{5} =$ $\frac{19}{6} =$ $\frac{17}{8} =$ $\frac{47}{9} =$ $\frac{33}{10} =$

Reduce each fraction to simplest form.

3. $\frac{10}{15} =$ $\frac{12}{18} =$ $\frac{9}{12} =$ $\frac{16}{40} =$ $\frac{25}{45} =$ $\frac{32}{24} =$ $\frac{52}{10} =$

Complete these sentences.

4. 6 inverted is _____. $\frac{2}{3}$ inverted is _____. $\frac{1}{2}$ inverted is _____. $\frac{9}{2}$ inverted is _____.

Divide.

5. $\frac{5}{8} \div 10 =$ $\frac{4}{5} \div 8 =$ $\frac{3}{5} \div 5 =$ $\frac{3}{4} \div 6 =$ $\frac{2}{3} \div 4 =$

6. $\frac{1}{3} \div \frac{1}{6} =$ $\frac{1}{6} \div \frac{1}{3} =$ $\frac{3}{5} \div \frac{9}{10} =$ $\frac{3}{8} \div \frac{3}{5} =$ $\frac{5}{8} \div \frac{3}{4} =$

7. $6 \div \frac{1}{2} =$ $10 \div \frac{2}{3} =$ $15 \div \frac{3}{5} =$ $2 \div \frac{4}{5} =$ $3 \div \frac{5}{8} =$

8. $3\frac{1}{3} \div 10 =$ $2\frac{1}{5} \div 4 =$ $5\frac{1}{4} \div 3 =$ $1\frac{4}{5} \div 9 =$ $4\frac{1}{2} \div 3 =$

9. $1\frac{7}{12} \div \frac{3}{4} =$ $3\frac{3}{4} \div \frac{1}{2} =$ $3\frac{2}{3} \div \frac{11}{12} =$ $1\frac{4}{7} \div \frac{1}{7} =$ $2\frac{2}{9} \div \frac{4}{5} =$

10. $8\frac{1}{3} \div 1\frac{1}{4} =$ $6\frac{2}{5} \div 5\frac{1}{3} =$ $1\frac{1}{3} \div 2\frac{2}{3} =$ $5\frac{2}{5} \div 1\frac{3}{5} =$ $1\frac{1}{5} \div 2\frac{4}{5} =$

Solve.

11. A farmer wants to divide a 10-acre field into $2\frac{1}{2}$ acre fields. How many fields would there be?

Answer _____

12. A grocer wants to stack 36 cases of fruit drinks in a display. She wants $4\frac{1}{2}$ cases in each stack. How many stacks will she have?

Answer _____

Solve. Choose the correct operation.   Do Your Work Here

1. Mr. Yanaga's car has a gas tank that holds 16 gallons. He used $\frac{3}{4}$ of a tank of gas. How many gallons of gas did he use?   _____

2. Sue and Mike bought 54 feet of lumber for building window frames. Each window frame requires $13\frac{1}{2}$ feet of lumber. How many window frames can they make?   _____

3. Terry lives $\frac{5}{8}$ mile from town. Jack lives halfway between Terry and town. How far from town does Jack live?   _____

4. The Sernas bought material to be used for making curtains. There are $24\frac{1}{2}$ yards of material. Each window requires $3\frac{1}{2}$ yards. How many sets of curtains can the Sernas make?   _____

5. ACME Contractors bought a 12-pound carton of glue. The glue was in $\frac{1}{8}$ pound tubes. How many tubes were there in the carton?   _____

6. A load of cement weighing $\frac{3}{4}$ ton was divided into bags, each weighing $\frac{1}{20}$ ton. How many bags did the load contain?   _____

7. Maria bought $25\frac{1}{2}$ feet of tubing for $2 per foot. How much did she pay for all of it?   _____

8. Lin wished to make T-shirts for each of her five children. Each shirt required $\frac{3}{4}$ yard of cloth. How many yards did she need in all?   _____

Perform the indicated operation.

1.

$\begin{array}{r} \frac{9}{10} \\ +\ \frac{2}{3} \\ \hline \end{array}$
$\qquad$
$\begin{array}{r} 5\frac{3}{5} \\ +3\frac{3}{4} \\ \hline \end{array}$
$\qquad$
$\begin{array}{r} \frac{3}{5} \\ -\ \frac{1}{10} \\ \hline \end{array}$
$\qquad$
$\begin{array}{r} 12\frac{5}{6} \\ -\ \frac{2}{5} \\ \hline \end{array}$
$\qquad$
$\begin{array}{r} 2\frac{2}{3} \\ +\ \frac{7}{8} \\ \hline \end{array}$

2.

$\begin{array}{r} \frac{7}{8} \\ +\ \frac{1}{2} \\ \hline \end{array}$
$\qquad$
$\begin{array}{r} 8\frac{2}{3} \\ -2\frac{2}{5} \\ \hline \end{array}$
$\qquad$
$\begin{array}{r} 10 \\ -\ \frac{3}{16} \\ \hline \end{array}$
$\qquad$
$\begin{array}{r} \frac{2}{5} \\ +6\frac{1}{4} \\ \hline \end{array}$
$\qquad$
$\begin{array}{r} 24\frac{1}{5} \\ -10\frac{2}{3} \\ \hline \end{array}$

3.

$\begin{array}{r} \frac{3}{4} \\ \frac{1}{6} \\ +\ \frac{1}{2} \\ \hline \end{array}$
$\qquad$
$\begin{array}{r} 8 \\ -\ \frac{7}{10} \\ \hline \end{array}$
$\qquad$
$\begin{array}{r} 13\frac{3}{4} \\ -\ \frac{2}{3} \\ \hline \end{array}$
$\qquad$
$\begin{array}{r} \frac{5}{8} \\ \frac{1}{4} \\ +\ \frac{1}{2} \\ \hline \end{array}$
$\qquad$
$\begin{array}{r} 6\frac{1}{9} \\ +4\frac{2}{3} \\ \hline \end{array}$

4. $\frac{3}{4} \times \frac{8}{9} =$ $\qquad$ $\frac{9}{10} \times \frac{2}{3} =$ $\qquad$ $3 \times \frac{1}{6} =$ $\qquad$ $\frac{3}{5} \times 25 =$

5. $2\frac{2}{3} \times 6 =$ $\qquad$ $3\frac{1}{2} \times \frac{1}{2} =$ $\qquad$ $2\frac{2}{5} \times 6\frac{1}{4} =$ $\qquad$ $1\frac{3}{10} \times 3\frac{1}{8} =$

6. $\frac{5}{6} \times \frac{3}{10} =$ $\qquad$ $2\frac{1}{2} \times 4 =$ $\qquad$ $3\frac{1}{3} \times 4\frac{3}{4} =$ $\qquad$ $\frac{7}{8} \times 3\frac{1}{2} =$

7. $\frac{7}{8} \div 14 =$ $\qquad$ $\frac{9}{10} \div \frac{3}{5} =$ $\qquad$ $17 \div \frac{1}{3} =$ $\qquad$ $3\frac{1}{3} \div 5 =$

8. $\frac{3}{4} \div \frac{1}{12} =$ $\qquad$ $\frac{5}{6} \div 15 =$ $\qquad$ $3\frac{1}{4} \div 2\frac{1}{2} =$ $\qquad$ $6 \div \frac{2}{3} =$

9. $5\frac{1}{3} \div \frac{1}{3} =$ $\qquad$ $12 \div 2\frac{2}{5} =$ $\qquad$ $3\frac{1}{3} \div 4\frac{1}{6} =$ $\qquad$ $4\frac{2}{3} \div 5\frac{1}{4} =$

Solve the following word problems.

10. Maria bought $25\frac{1}{2}$ yards of red ribbon, $32\frac{3}{4}$ yards of blue ribbon, and $40\frac{1}{4}$ yards of yellow ribbon. How much ribbon did she buy?

Answer _____

11. From New Orleans to El Paso is 1,224 miles. The fast train makes this trip in $22\frac{2}{3}$ hours. What is the average speed?

Answer _____

12. Meili bought $3\frac{1}{2}$ dozen eggs. She used $\frac{1}{3}$ dozen. How many dozen eggs are left?

Answer _____

13. A ream of paper is $2\frac{1}{4}$ in. thick. How many reams will fit on a shelf 9 in. high?

Answer _____

Unit 3 Decimals

The Meaning of Decimals

Like fractions, decimals show parts of a whole. The shaded portion of each picture can be written as a fraction or as a decimal.

 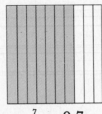 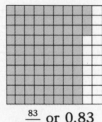 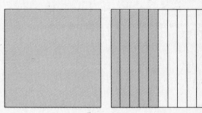

$\frac{1}{1}$ or 1 $\frac{7}{10}$ or 0.7 $\frac{83}{100}$ or 0.83 $1\frac{5}{10}$ or 1.5

Read: one   seven tenths   eighty-three hundredths   one and five tenths

Remember,
- a decimal point separates a whole number and its decimal parts.
- a whole number has a decimal point but it is usually not written. For example, 2 = 2.0 and $9 = $9.00.
- a decimal point is read as "and."
- a zero in front of a decimal is a placeholder. For example, $0.25 = $.25 and 0.36 = .36.

Write the decimal shown by the shaded part of each figure.

1.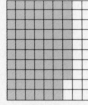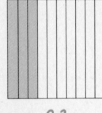

   _____0.3_____   _____   _____

2.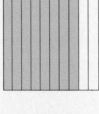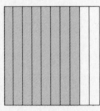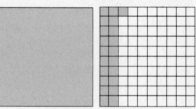

   _____   _____   _____

Write each money amount with a dollar sign and a decimal point.

3. one dollar ___$1.00___   ten cents ___$0.10___   one penny ___$0.01___

4. twelve dollars _____   three dimes _____   eight pennies _____

5. five dollars and eleven   thirty-seven cents _____   sixty-four cents _____

   cents _____

76

Reading and Writing Decimals

To read a decimal, read as a whole number. Then name the place value of the last digit.

Read and write 0.53 as fifty-three hundredths.

To read a decimal that has a whole number part,

- read the whole number part.
- read the decimal point as "and."
- read the decimal part as a whole number and then name the place value of the last digit.

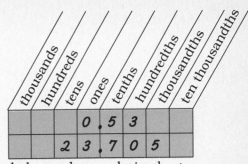

← whole number . decimal →

Read and write 23.705 as twenty-three and seven hundred five thousandths.

Read and write the following as decimals.

1. three tenths ___0.3___ 15 ten thousandths _____

2. 25 thousandths _____ five thousandths _____

3. 15 hundredths _____ three-tenths foot _____

4. five tenths of a yard _____ seventy-five hundredths of a mile _____

5. 15 thousandths of an inch _____ 4 hundredths of a mile _____

6. 25 hundredths of a ton _____ 25 ten thousandths _____

Write the following decimals in word form.

7. A dime is 0.1 of a dollar ___one tenth_____

8. A cent is 0.01 of a dollar _____

9. A mill is 0.1 of a cent _____

10. A quarter is 0.25 of a dollar _____

11. A half-dollar is 0.5 or 0.50 of a dollar _____

12. A gallon of water weighs 8.23 lb _____

13. A gallon of milk averages 8.59 lb _____

14. A knot is equal to 1.1516 miles _____

15. A meter is 39.37 inches _____

16. A kilometer equals 0.621 mile _____

Making use of the illustrations in the box above, read these decimals.

17. 0.5   0.002   1,892.105   3,456.101

18. 0.05   0.012   2,236.1   2,987.17

19. 0.005   125.15   1,594.001   3,642.9

Comparing and Ordering Decimals

To compare two decimal numbers, begin at the left. Compare the digits in each place.

The symbol < means "is less than." $4.2 < 4.6$
The symbol > means "is greater than." $2.7 > 2.3$
The symbol = means "is equal to." $3.4 = 3.40$

Compare: 2.6 and 2.3

| 2 . 6 |
| 2 . 3 |

The ones digits are the same. Compare the tenths.

$6 > 3$, so $2.6 > 2.3$

Compare: 0.08 and 0.25

| 0 . 0 8 |
| 0 . 2 5 |

The ones digits are the same. Compare the tenths.

$0 < 2$, so $0.08 < 0.25$

Compare: 0.4 and 0.47

| 0 . 4 0 |
| 0 . 4 7 |

Write a zero. The ones and tenths digits are the same. Compare the hundredths.

$0 < 7$, so $0.4 < 0.47$

Compare. Write <, >, or =.

1. 0.3 __<__ 0.32   0.035 _____ 0.35   0.87 _____ 0.087   0.5 _____ 0.500

0 . 3 0
0 . 3 2

2. 0.135 _____ 0.14   0.15 _____ 0.115   9.99 _____ 9.09   3.2 _____ 3.20

Compare the numbers in each pair. Draw a line under the one which is larger. If the two numbers are equal, draw a line under both.

3. 0.3 and 0.30   0.5 and 0.500   0.053 and 0.53   0.25 and 0.3

4. 3.5 and 3.50   4.50 and 4.500   0.125 and 13   0.51 and 0.151

5. 0.625 and 0.6250   0.035 and 0.0350   0.26 and 0.3   0.6 and 0.65

Write these numbers in order, beginning with the smallest.

6. 25, 2.5, 0.25, 1.25, 1.025, 1.20, 1.1, 1.01 _____

When the numbers in each pair below are the same, write S on the line.
When they are not the same, write D on the line.

7. One hundred twenty-five — 0.125 __D__   Three fourths — 0.34 _____

8. One and six tenths — 1.6 _____   Four twenty-fifths — 4.25 _____

9. Thirty-five and one half — 35.2 _____   Eight thousandths — 0.008 _____

10. Ten and one fifth — 10.5 _____   Five and one tenth — 5.10 _____

Fraction and Decimal Equivalents

Sometimes you will need to either change a decimal to a fraction or a fraction to a decimal. Notice that the number of decimal places in the decimal is the same as the number of zeros in the denominator.

one tenth = $\frac{1}{10}$ = 0.1	one hundredth = $\frac{1}{100}$ = 0.01
five tenths = $\frac{5}{10}$ = 0.5	twelve hundredths = $\frac{12}{100}$ = 0.12
one and six tenths = $1\frac{6}{10}$ = 1.6	one thousandth = $\frac{1}{1000}$ = 0.001
six and fifteen hundredths = $6\frac{15}{100}$ = 6.15	fifteen thousandths = $\frac{15}{1000}$ = 0.015

In each column, write the equivalent numbers in words, fractions, or decimals.

	COLUMN A	COLUMN B	COLUMN C
1.	three tenths	$\frac{3}{10}$	0.3
2.	fifteen hundredths		
3.		$\frac{5}{1000}$	
4.			0.027
5.			4.6
6.		$15\frac{7}{10}$	
7.	thirty and three hundredths		
8.	one hundred twenty and two thousandths		

Not all fractions can be changed to decimal form easily. To write fractions that have denominators other than 10, 100, or 1,000 as decimals, first write an equivalent fraction that has a denominator of 10, 100, or 1,000. Then write the equivalent fraction as a decimal.

Write $\frac{1}{5}$ as a decimal.

Write $\frac{1}{5}$ with 10 as the denominator.	Write the fraction as a decimal.
$\frac{1}{5} = \frac{1 \times 2}{5 \times 2} = \frac{2}{10}$	= 0.2

Write $2\frac{3}{4}$ as a decimal.

Write $2\frac{3}{4}$ as an improper fraction.	Write the new fraction with 100 as the denominator.	Write the fraction as a decimal.
$2\frac{3}{4} = \frac{11}{4}$	$\frac{11}{4} = \frac{11 \times 25}{4 \times 25} = \frac{275}{100}$	= 2.75

Change each fraction to an equivalent fraction and then to a decimal.

9. $\frac{1}{5} = \frac{1 \times 2}{5 \times 2} = \frac{2}{10} = 0.2$ $\frac{3}{5} =$ $\frac{4}{5} =$ $\frac{1}{2} =$

10. $4\frac{4}{25} = \frac{104 \times 4}{25 \times 4} = \frac{416}{100} = 4.16$ $2\frac{7}{20} =$ $1\frac{9}{25} =$ $3\frac{42}{50} =$

Rounding Decimals

Rounding decimals can be used to tell about how many. You can use a number line to round decimals.

Remember, when a number is halfway, always round up.

Round 31.2 to the nearest one.

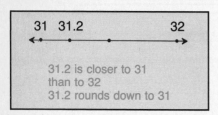

31.2 is closer to 31 than to 32
31.2 rounds down to 31

Round $4.67 to the nearest dollar.

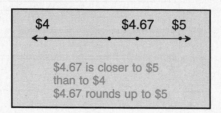

$4.67 is closer to $5 than to $4
$4.67 rounds up to $5

Round 6.15 to the nearest tenth.

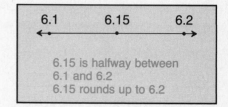

6.15 is halfway between 6.1 and 6.2
6.15 rounds up to 6.2

Round to the nearest one.

1. 4.4 ___4___   3.6 _____   2.5 _____   8.4 _____   7.3 _____

2. 49.7 __50__   51.5 _____   44.6 _____   79.1 _____   55.5 _____

3. 59.5 _____   99.4 _____   67.3 _____   40.7 _____   79.12 _____

4. 6.39 ___6___   8.76 _____   5.02 _____   9.93 _____   3.11 _____

5. 2.71 _____   9.79 _____   8.65 _____   1.01 _____   7.77 _____

Round each amount to the nearest dollar.

6. $3.92 __$4__ $5.47 _____ $7.92 _____ $6.35 _____ $7.81 _____

7. $8.04 _____ $2.56 _____ $9.53 _____ $6.06 _____ $7.77 _____

8. $21.63 __$22__ $11.58 _____ $20.30 _____ $19.99 _____ $62.62 _____

9. $1.21 _____ $6.49 _____ $2.95 _____ $8.50 _____ $9.09 _____

10. $16.17 _____ $33.33 _____ $70.07 _____ $99.10 _____ $29.89 _____

Round to the nearest tenth.

11. 0.58 __0.6__   0.91 _____   0.64 _____   0.79 _____   0.09 _____

12. 4.08 __4.1__   8.67 _____   2.34 _____   9.33 _____   8.88 _____

13. 61.97 __62.0__   47.96 _____   99.99 _____   50.95 _____   48.97 _____

14. 39.96 _____   25.81 _____   72.02 _____   21.63 _____   32.74 _____

15. 40.94 _____   30.89 _____   55.55 _____   11.64 _____   19.63 _____

Adding Decimals

To add decimals, line up the decimal points. Write zeros as needed. Then add as with whole numbers. Be sure to write a decimal point in the sum.

Find: 4.7 + 7.32

Write a zero.	Add the hundredths.	Add the tenths. Regroup. Write a decimal point in the sum.	Add the ones.
T O Ts Hs	T O Ts Hs	T O Ts Hs	T O Ts Hs
4 .7 0 ↙	4 .7 **0**	¹ 4 .7 0	¹ 4 .7 0
+ 7 .3 2	+ 7 .3 **2**	+ 7 .3 2	+ 7 .3 2
	2	.0 2	1 2 .0 2

Add. Write zeros as needed.

1.
```
    1                                                                        
  5.9 0↙      7.0 8      1 4.0 7 6     1 7.0 5      4 2.0 8      3 9.9
+ 3.6 2    + 3.2 6 5    +   8.4 6    +   3.3 5 1   +   0.1 6    + 3 9.
  9.5 2
```

2.
```
  1     1
2 8.0 0 9↙    7 7.0 1 6     8 4.7        8.2 3      0.3 5 6      7.1 1 1
  4.6 5 0↙      4.5 7       0.4 0 3     3 6.5        4.3 0      9.1 1
+   6.0 0 3   +   0.6 4 7   + 3.0 8    + 0.0 0 9    + 0.8 1 7   + 6.2 0 3
3 8.6 6 2
```

3.
```
    1
1 2.2 0↙      8.0 5 0      0.2 2        6.1 2 6      9.4 5 1     1 9.0 0 3
  2.1 0↙    1 4.0 0 5      0.3 5 6    1 4.0 0 5      0.0 0 7      0.1
  4.0 0↙    1 6.1          0.4 3 9      0.0 4        7.3        1 2.0 0 4
+   5.2 5   +   7.3 2      + 0.8      +   2.6      + 6.5 3 6    +   5.5 5
2 3.5 5
```

4.
```
  2.0 0 5↙     0.0 0 9      0.1 4        0.7 2        6.1 1 2      9 9.1 1 1
  0.1 5 0↙     0.1 4        0.0 6 0      0.0 0 5      4.2 2          6.7
  0.6 0 0↙     0.6         0.1 7         0.0 3      2 4.7 3 6      3 2.9 9
+ 3.9 1 0↙   + 2.1 0       + 8.5       + 7.1 6     + 1 4.1        +   7.7 7
```

Line up the decimal points. Then add. Write zeros as needed.

5. 9 + 3.4 + 0.7 = _____   6.54 + 10 + 8.35 = _____   3 + 4.51 + 7.3 = _____

```
  9.0↙
  3.4
+ 0.7
```

6. 8 + 9.5 + 0.1 = _____   9.99 + 8 + 5.75 = _____   9 + 11.1 + 7.62 = _____

Subtracting Decimals

To subtract decimals, line up the decimal points. Write zeros as needed. Then subtract as with whole numbers. Be sure to write a decimal point in your answer.

Find: 34.3 − 17.94

Write a zero. Regroup to subtract the hundredths.	Regroup to subtract the tenths. Write a decimal point in the difference.	Regroup to subtract the ones.	Subtract the tens.
T O Ts Hs	T O Ts Hs	T O Ts Hs	T O Ts Hs

3 4 .3 $\overset{2}{\cancel{3}}\overset{10}{\cancel{0}}$
 − 1 7 .9 4
            6
	3 4 .$\overset{12}{\cancel{3}}\overset{2}{}\overset{10}{\cancel{0}}$
	 − 1 7 .9 4
	       .3 6
		3 $\overset{13}{\cancel{4}}$.$\overset{12}{\cancel{3}}$ $\overset{10}{\cancel{0}}$
		 − 1 7 .9 4
		     6 .3 6
			$\overset{2}{\cancel{3}}$ $\overset{13}{\cancel{4}}$.$\overset{12}{\cancel{3}}$ $\overset{10}{\cancel{0}}$
			 − 1 7 .9 4
			    1 6 .3 6

Subtract. Write zeros as needed.

1.

8.3 2 $\overset{1}{}\overset{10}{\cancel{0}}$ − 3.2 0 3 5.1 1 7	7.2 1 8 − 5.1 2	5.0 9 1 − 3.6 2	9.0 6 − 7.0 5 4	2.7 4 9 − 1.1 5	1 0.3 9 9 − 1 0.2 3

2.

3 4.9 5 − 2 7.9 9	9 2.0 0 − 6 7.5 0	6 3.5 5 − 3 0.7 0	9 4.7 8 − 1 5.0 0	2 6.7 7 − 5.6 8	3 2.3 0 − 2 0.0 0

3.

$\overset{5}{}\overset{11}{\cancel{6}}$ .2 $\overset{9}{\cancel{0}}\overset{10}{\cancel{0}}$ − 4.5 7 5 1.6 2 5	1.9 − 0.6 7 4	3.6 − 0.7 9 1	1.3 5 4 − 0.2 6	7.3 − 0.9 1 2	9 2.1 5 − 8 4.7

4.

8.0 9 − 4.2 5 6	9 4.7 8 − 1 5.1	7.3 0 − 5.0 0 2	1 9.0 0 5 − 1 4.5	3.0 7 2 − 2.1	1.5 2 − 0.4 0 8

Line up the decimal points. Then subtract. Write zeros as needed.

5. 7.05 − 3.035 _____   8.14 − 6.1 _____   75.06 − 52.8 _____

     7.0 5 0
  − 3.0 3 5

Solve these word problems.

6. During a snowstorm, 8.5 in. of snow fell in Chicago. In Detroit only 3.8 in. fell. How much more snow fell in Chicago?

Answer _____

7. A machinist found a rod to be 2.15 inches across. She cut it down to 1.92 inches. How much did she cut?

Answer _____

Adding and Subtracting Money

When adding or subtracting money, write a decimal point and a dollar sign in your answer. Be sure to line up your decimal points.

Find: $0.39 + $0.47

Add the hundredths (pennies) column. Regroup.	Add the tenths (dimes) column. Write a decimal point. Write a dollar sign.

O	Ts	Hs
	1	
$0	.3	9
+ 0	.4	7
		6

O	Ts	Hs
	1	
$0	.3	9
+ 0	.4	7
$0	.8	6

Find: $0.43 − $0.26

Regroup to subtract the hundredths (pennies) column.	Subtract the tenths (dimes) column. Write a decimal point. Write a dollar sign.

O	Ts	Hs
	3	*13*
$0	.4̷	3̷
− 0	.2	6
		7

O	Ts	Hs
	3	*13*
$0	.4̷	3̷
− 0	.2	6
$0	.1	7

Rewrite the following amounts of change.

1. $0.28 __1__ quarter and __3__ pennies
2. $0.15 _____ dime and __1__ nickel
3. $0.16 __1__ dime and _____ pennies
4. $0.23 _____ dimes and __3__ pennies
5. $0.19 _____ nickels and __4__ pennies
6. $0.55 __5__ dimes and _____ nickel
7. $0.78 _____ quarters and __3__ pennies
8. $0.27 _____ dimes and __7__ pennies
9. $0.26 _____ nickels and __1__ penny
10. $0.95 _____ dimes and __1__ nickel

Add. Write a decimal point and a dollar sign in your answer.

11.
$0.25 + 0.38 = $0.63
$0.54 + 0.29
$0.28 + 0.36
$0.35 + 0.49
$0.37 + 0.46
$0.29 + 0.52
$0.30 + 0.54

Subtract.

12.
$0.29 − 0.21 = $0.08
$0.45 − 0.20
$0.32 − 0.18
$0.30 − 0.24
$0.29 − 0.26
$0.34 − 0.19
$0.26 − 0.24

Add or subtract.

13.
$0.40 + 0.68
$0.46 + 0.77
$0.74 − 0.69
$0.33 + 0.88
$0.55 − 0.45
$0.60 − 0.45
$0.67 − 0.54

Solve.

Do Your Work Here

14. Tanya has four quarters and three dimes. Juan has three quarters and five pennies. How much do they have in all? _____

Adding and Subtracting Money

Study the examples which show how to add and subtract money.
Remember to add the decimal point and dollar sign to your answers.

Find: $2.25 + $3.15

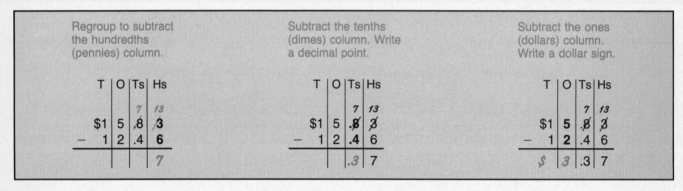

Add the hundredths (pennies) column. Regroup.	Add the tenths (dimes) column. Write a decimal point.	Add the ones (dollars) column. Write a dollar sign.
O | Ts | Hs	O | Ts | Hs	O | Ts | Hs
$2 .2 5 + 3 .1 5 ——— 0	$2 .2 5 + 3 .1 5 ——— .4 0	$2 .2 5 + 3 .1 5 ——— $5 .4 0

Find: $15.83 − $12.46

Regroup to subtract the hundredths (pennies) column.	Subtract the tenths (dimes) column. Write a decimal point.	Subtract the ones (dollars) column. Write a dollar sign.
T | O | Ts | Hs	T | O | Ts | Hs	T | O | Ts | Hs
$1 5 .8 3 − 1 2 .4 6 ——— 7	$1 5 .8 3 − 1 2 .4 6 ——— .3 7	$1 5 .8 3 − 1 2 .4 6 ——— $ 3 .3 7

Add.

1.
   $2.19 $2.15 $2.26 $4.25 $2.26 $12.15 $12.25
 + 5.14   + 1.49   + 3.17   + 3.35   + 3.15   + 13.48   + 11.47
 ———————
  $7.33

Subtract.

2.
   $5.46 $7.54 $2.50 $4.76 $8.34 $13.97 $17.54
 − 2.39   − 6.38   − 1.24   − 2.38   − 6.17   − 12.29   − 13.38
 ———————
  $3.07

Add or subtract.

3.
   $3.19 $5.72 $4.90 $2.48 $3.01 $28.17 $23.46
 + 2.22   − 2.15   − 1.36   + 4.33   + 2.49   − 15.09   + 11.27

Solve.

4. Patricia bought a jacket for $65.15 and shoes for $43.45. How much did she spend?

Answer _____

5. Mitchell bought a shirt for $14.15 and a belt for $9.95. How much did he spend?

Answer _____

Estimating Decimal Sums and Differences

To estimate decimal sums, first round the decimals to the same place. Then add the rounded numbers.

Estimate: $7.69 + $4.19

Round each decimal to the nearest dollar. Add.

$$\begin{array}{rcr} \$7.6\,9 & \to & \$\ \ 8 \\ +\ \ 4.1\,9 & \to & +\ \ \ 4 \\ \hline & & \$1\,2 \end{array}$$

To estimate decimal differences, first round the decimals to the same place. Then subtract the rounded numbers.

Estimate: 10.34 − 6.78

Round each decimal to the nearest tenth. Subtract.

$$\begin{array}{rcr} 1\,0.3\,4 & \to & 1\,0.3 \\ -\ \ 6.7\,8 & \to & -\ \ 6.8 \\ \hline & & 3.5 \end{array}$$

Estimate the sum or difference by rounding to the nearest dollar.

1.
$$\begin{array}{rcr} \$7.6\,5 & \to & \$\ \ 8 \\ +\ \ 5.3\,3 & \to & +\ \ \ 5 \\ \hline & & \$1\,3 \end{array}$$

$$\begin{array}{rcl} \$1\,0.4\,5 & \to & \\ +\ \ 2\,3.5\,6 & \to & \\ \hline \end{array}$$

$$\begin{array}{rcl} \$\ \ 9\,9.9\,0 & \to & \\ +\ \ 1\,2\,1.2\,5 & \to & \\ \hline \end{array}$$

$$\begin{array}{rcl} \$2\,6.7\,6 & \to & \\ +\ \ \ \ 1.9\,8 & \to & \\ \hline \end{array}$$

2.
$$\begin{array}{rcr} \$9\,9.7\,6 & \to & \$1\,0\,0 \\ -\ \ 2\,0.3\,0 & \to & -\ \ \ 2\,0 \\ \hline & & \$\ \ 8\,0 \end{array}$$

$$\begin{array}{rcl} \$1\,0.4\,5 & \to & \\ -\ \ \ 9.2\,3 & \to & \\ \hline \end{array}$$

$$\begin{array}{rcl} \$9.8\,7 & \to & \\ -\ \ 5.5\,7 & \to & \\ \hline \end{array}$$

$$\begin{array}{rcl} \$1\,3.4\,5 & \to & \\ -\ \ 1\,2.8\,9 & \to & \\ \hline \end{array}$$

3.
$$\begin{array}{rcl} \$8.5\,4 & \to & \\ -\ \ 4.5\,0 & \to & \\ \hline \end{array}$$

$$\begin{array}{rcl} \$3\,0.6\,8 & \to & \\ +\ \ 2\,4.1\,0 & \to & \\ \hline \end{array}$$

$$\begin{array}{rcl} \$\ \ 2\,3.0\,5 & \to & \\ +\ \ 1\,0\,1.8\,6 & \to & \\ \hline \end{array}$$

$$\begin{array}{rcl} \$2\,9.7\,0 & \to & \\ -\ \ \ \ 8.2\,9 & \to & \\ \hline \end{array}$$

Estimate the sum or difference by rounding to the nearest tenth.

4. 12.35 − 2.17

$$\begin{array}{rcr} 1\,2.3\,5 & \to & 1\,2.4 \\ -\ \ 2.1\,7 & \to & -\ \ 2.2 \\ \hline & & 1\,0.2 \end{array}$$

5.08 + 3.07

19.18 − 9.28

5. 10.67 + 9.33

4.75 − 0.66

300.31 + 25.32

6. 25.04 + 4.86

30.28 − 9.83

175.37 + 24.62

Checking Up

Change each fraction to a decimal.

1. $\frac{1}{10}$ = _____ $\frac{3}{10}$ = _____ $\frac{1}{5}$ = _____ $\frac{1}{2}$ = _____ $\frac{1}{4}$ = _____ $\frac{3}{4}$ = _____

Compare. Write <, >, or =.

2. 0.5 _____ 0.52   0.117 _____ 0.17   0.13 _____ 0.133

3. 0.45 _____ 4.50   0.6 _____ 0.600   0.22 _____ 2.026

Round each amount to the nearest tenth.

4. 3.87 _____   4.78 _____   5.21 _____   6.34 _____   9.75 _____

Round each amount to the nearest dollar.

5. $7.98 _____ $8.46 _____ $7.33 _____ $9.18 _____ $6.50 _____

Write each number as a decimal.

6. Six thousandths _____   Twelve hundredths _____

7. Twenty thousandths _____   Four hundredths _____

8. Four tenths _____   Forty thousandths _____

In each column write the equivalent decimal or fraction in words, fractions, or decimals.

9. ___ four tenths ___   _____   _____

10. _____ $\frac{6}{1000}$ _____

11. _____   _____   3.7

Add.

12.
2.005	$5.07	$7.14	0.702	17.005	31.09
+ 0.15	+ 6.14	+ 0.06	+ 0.0050	20.25	20.036
				+ 16	50.17
					+ 12.4

Subtract.

13.
7.54	5.498	$7.51	6.2	$8.90	1.9
− 6.38	− 2.362	− 2.28	− 4.57	− 4.25	− 0.674

Estimate the sum or difference by rounding to the nearest tenth.

14. 36.08 + 5.92   40.36 − 8.79   186.43 + 34.58

Problem-Solving Strategy: Use Estimation

Many problems can be solved by estimation. Often, you do not need an exact answer to solve a problem. An estimate is found by rounding some or all of the numbers and then doing mental math. An estimate can help you decide if an answer is reasonable.

STEPS

1. **Read the problem.**
   Green Things Plant Shop was having a sale. Tulips were on sale for $0.45 each. Cacti were on sale for $0.79 each. Rose bushes were on sale for $2.99 each.
   Joe has $5.00. If he buys one rose bush, can he also buy two cactus plants?

2. **Identify the important facts.**
   Joe has $5.00.
   One rose bush costs $2.99.
   One cactus plant costs $0.79.

3. **Round.**
   Round $2.99 to the nearest dollar.
   $2.99 rounds to $3.
   Round $0.79 to the nearest tenth of a dollar (dime).
   $0.79 rounds to $0.80.

4. **Solve the problem.**
   Think: $3.00 for a rose bush and 2 × $0.80 = $1.60 for the cactus.
   $3.00 + $1.60 = $4.60
   Joe has enough money because $4.60 < $5.00.

Use estimation to solve each problem. Round to the nearest dollar.

1. Polly wants to buy 2 pounds of T-bone steak at $3.79 per pound. About how much money does she need?

   Answer _____

2. Giorgio has $20.00. Can he buy 3 pounds of sirloin steak at $2.39 per pound and 4 pounds of round steak at $1.99 per pound?

   Answer _____

3. James bought 3 pounds of round steak at $1.99 per pound and 4 pounds of chuck steak at $1.89 per pound. About how much money did he get back from his 20-dollar bill?

   Answer _____

4. T-bone steak costs $3.79 per pound. About how many pounds of T-bone steak can you buy for $29?

   Answer _____

Solve.

1. Denver has an elevation of 5,183.4 feet. El Paso has an elevation of 3,720.7 feet. How much greater is the Denver elevation?

    Answer _____

2. Rosa delivers packages. Rosa drove to Sandoval, a distance of 8.2 kilometers. Then she drove 16.5 kilometers to Richview. From there, she drove 14.9 kilometers to Mount Vernon. Finally she drove 39.9 kilometers to Marion. How far did Rosa drive in all?

    Answer _____

3. Mr. Ford bought a rug at a sale price of $167.50. The original price was $192. How much did he save?

    Answer _____

4. Los Angeles is 482.5 miles from San Francisco. Fresno is located between these two cities and is 205.7 miles from San Francisco. How far is it from Fresno to Los Angeles?

    Answer _____

5. Jan is painting her kitchen. She bought a gallon of paint for $11.47, a brush for $2.49, masking tape for $0.59, and a package of sandpaper for $1.77. About how much did Jan pay for the painting supplies? Use estimation. Round to the nearest dollar.

    Answer _____

6. Pears cost $1.49 per pound. About how many pounds of pears can you buy for $5.00?

    Answer _____

7. From New Orleans to El Paso it is 1,975.7 kilometers. Tucson is 500.5 kilometers farther. How far is it from New Orleans to Tucson through El Paso?

    Answer _____

8. Connie was comparing shoe prices. She found one pair that cost $36.95. She found a similar pair that cost $1.92 more. What was the price of the second pair?

    Answer _____

9. Sarah Jennings made a trip by plane from Seattle to Houston, a distance of 2,449.67 miles. David Marshall made a trip by car from Los Angeles to Minneapolis, a distance of 2,018.7 miles. How many miles did they travel in all?

    Answer _____

10. Ho bought four items at the flea market. The prices were $3.25, $6.99, $2.39, and $2.75. About how much money did he get back from his 20-dollar bill? Use estimation. Round to the nearest dollar.

    Answer _____

11. One automobile piston has a diameter of 4.005 inches. Another piston has a diameter of 3.995 inches. How much larger is the first piston?

    Answer _____

12. One experimental train went 120 miles an hour. Another experimental train went 112.5 miles an hour. How much difference is there between these two speeds? (Hint: 120 = 120.0)

    Answer _____

Multiplying Decimals by Whole Numbers

To multiply decimals by whole numbers, multiply as if you were multiplying whole numbers. Count the number of decimal places to the right of the decimal point in the numbers you have multiplied. The product will have the same number of decimal places. Place the decimal point in the product.

Remember, sometimes you might need to write a zero in the product in order to place the decimal point correctly.

Find: 13 × 6.4

6.4	1 decimal place
× 13	+ 0 decimal places
192	
64	
83.2	1 decimal place

Find: 0.018 × 5

5	0 decimal places
× 0.018	+ 3 decimal places
0.090	3 decimal places
	Write a zero.

Place the decimal point in the product.

1.
0.2	2	0.3	5	5	94	2.18	0.707
× 2	× 0.2	× 4	× 0.5	× 9	× 0.4	× 0.3	× 0.2
04	*04*	*12*	*25*	*45*	*376*	*654*	*1414*

2.
0.28	17	1.2	0.84	0.35	24	1.6	0.28
× 5	× 0.5	× 4	× 5	× 4	× 0.20	× 40	× 30
140	*85*	*48*	*420*	*140*	*480*	*640*	*840*

Multiply. Write zeros as needed.

3.
0.862	0.084	1.63	2.34	13.6	28.52	15.2
× 2	× 3	× 6	× 5	× 3	× 4	× 6
1.724						

4.
26	15	0.48	17	75	2.6	0.831
× 0.7	× 0.6	× 7	× 0.9	× 0.5	× 8	× 3
18.2						

5.
1.3	26	0.16	35	8.2	62	70
× 13	× 1.3	× 14	× 0.15	× 12	× 0.35	× 5.0
39						
130						
16.9						

6.
24	0.707	5	600	3	111	0.047
× 0.007	× 12	× 0.044	× 0.03	× 0.017	× 0.5	× 22
0.168						

Multiplying Decimals by Decimals

To multiply decimals by decimals, multiply as if you were multiplying whole numbers. Place the decimal point in the product by counting the number of decimal places to the right of the decimal point in both numbers. The product will have the same number of decimal places. Write zeros as needed.

Find: 0.48 × 13.7

Multiply. Write the decimal point in the product.

$$
\begin{array}{r}
\textbf{1 3.7} \quad \text{1 place} \\
\times\,\textbf{0.4 8} \quad \text{+ 2 places} \\
\hline
1096 \\
5480 \\
\hline
6.576 \quad \text{3 places}
\end{array}
$$

Find: 0.008 × 0.137

Multiply. Write the decimal point in the product.

$$
\begin{array}{r}
\textbf{0.1 3 7} \quad \text{3 places} \\
\times\,\textbf{0.0 0 8} \quad \text{+ 3 places} \\
\hline
0.001096 \quad \text{6 places}
\end{array}
$$

Write 2 zeros.

Multiply. Write zeros if needed.

1.
$\begin{array}{r} 0.3 \\ \times\,0.6 \\ \hline 0.18 \end{array}$
$\quad$
$\begin{array}{r} 0.0\,3 \\ \times\quad 0.6 \\ \hline \end{array}$
$\quad$
$\begin{array}{r} 9.8 \\ \times\,0.5 \\ \hline \end{array}$
$\quad$
$\begin{array}{r} 6.3 \\ \times\,0.0\,4 \\ \hline \end{array}$
$\quad$
$\begin{array}{r} 1.6 \\ \times\,0.4 \\ \hline \end{array}$
$\quad$
$\begin{array}{r} 5.3 \\ \times\,0.0\,9 \\ \hline \end{array}$
$\quad$
$\begin{array}{r} 0.7\,6 \\ \times\quad 0.5 \\ \hline \end{array}$

2.
$\begin{array}{r} 0.1\,8\,4 \\ \times\quad 0.0\,7 \\ \hline 0.01288 \end{array}$
$\quad$
$\begin{array}{r} 2.0\,4 \\ \times\quad 0.2 \\ \hline \end{array}$
$\quad$
$\begin{array}{r} 5.1\,9 \\ \times\,0.0\,3 \\ \hline \end{array}$
$\quad$
$\begin{array}{r} 0.1\,4\,2 \\ \times\,0.0\,0\,7 \\ \hline \end{array}$
$\quad$
$\begin{array}{r} 0.1\,5\,5 \\ \times\,0.0\,0\,1 \\ \hline \end{array}$

3.
$\begin{array}{r} 2.5 \\ \times\,5.7 \\ \hline \end{array}$
$\quad$
$\begin{array}{r} 3.4\,7 \\ \times\quad 1.4 \\ \hline \end{array}$
$\quad$
$\begin{array}{r} 1\,6.5 \\ \times\quad 2.8 \\ \hline \end{array}$
$\quad$
$\begin{array}{r} 7.0\,1 \\ \times\,0.0\,3\,3 \\ \hline \end{array}$
$\quad$
$\begin{array}{r} 0.9\,8\,1 \\ \times\,0.0\,3\,4 \\ \hline \end{array}$

4.
$\begin{array}{r} 7.0\,5 \\ \times\,2.0\,4 \\ \hline 2820 \\ 0000 \\ 141000 \\ \hline 14.3820 \end{array}$
$\quad$
$\begin{array}{r} 1.5\,5 \\ \times\,3.1\,1 \\ \hline \end{array}$
$\quad$
$\begin{array}{r} 0.7\,5\,1 \\ \times\quad 3.0\,1 \\ \hline \end{array}$
$\quad$
$\begin{array}{r} 6\,6.6 \\ \times\,0.1\,2\,3 \\ \hline \end{array}$
$\quad$
$\begin{array}{r} 9\,0.1 \\ \times\,7\,0.1 \\ \hline \end{array}$

Line up the digits. Then multiply. Write zeros as needed.

5. 0.43 × 0.02 = _____
$\qquad$ 0.206 × 0.37 = _____
$\qquad$ 8.79 × 6.08 = _____

$\begin{array}{r} 0.4\,3 \\ \times\,0.0\,2 \\ \hline \end{array}$

6. 3.44 × 0.7 = _____
$\qquad$ 2.91 × 0.07 = _____
$\qquad$ 6.66 × 0.012 = _____

Multiplying by 10, 100, 1,000

To multiply decimals by powers of ten, move the decimal point in the product to the right as many places as there are zeros in the multiplier.

Remember, sometimes you might need to write zeros in the product in order to move the decimal point the correct number of places.

Study these examples.

$10 \times 0.89 = 8.9$ $100 \times 0.73 = 73$ $1,000 \times 0.52 = 520$

$10 \times 8.9 = 89$ $100 \times 7.3 = 730$ $1,000 \times 5.2 = 5,200$

Multiply. Write zeros as needed.

1. $7.5 \times 10 =$ ___75___ $46 \times 10 =$ _____ $0.07 \times 10 =$ _____

2. $100 \times 0.7 =$ _____ $100 \times 4.6 =$ _____ $0.075 \times 100 =$ _____

3. $0.5 \times 10 =$ ___5___ $8 \times 1,000 =$ _____ $1.25 \times 100 =$ _____

4. $12.5 \times 100 =$ _____ $0.125 \times 1,000 =$ _____ $14.92 \times 100 =$ _____

5. $6.2 \times 1,000 =$ _____ $642.15 \times 10 =$ _____ $642.15 \times 100 =$ _____

6. $3.15 \times 1,000 =$ _____ $0.048 \times 1,000 =$ _____ $0.048 \times 100 =$ _____

7. $0.375 \times 10 =$ _____ $3.75 \times 10 =$ _____ $37.5 \times 10 =$ _____

8. $375 \times 10 =$ _____ $0.375 \times 1,000 =$ _____ $0.007 \times 1,000 =$ _____

9. $719.35 \times 100 =$ _____ $16.147 \times 1,000 =$ _____ $14.92 \times 1,000 =$ _____

10. $267.18 \times 100 =$ _____ $2.6718 \times 100 =$ _____ $2.6718 \times 1,000 =$ _____

Solve.   ▓▓▓▓▓▓ **Do Your Work Here** ▓▓▓▓▓▓

11. What will be the weight of 100 gallons of water, since 1 gallon weighs 8.355 pounds?   _____

12. How much will 1,000 gallons of milk weigh if 1 gallon weighs 8.605 pounds?   _____

13. If our car uses 0.7 gallon of gas in going 10 miles, how much will it use on a 100-mile trip? (How many tens are there in 100?)   _____

14. If a racing car goes 1 mile in 0.4 minute, how long will it require to go 60 miles? (60 is a multiple of 10. Multiply by 6 and move the decimal point 1 place to the right.)   _____

Dividing Decimals by Whole Numbers

To divide a decimal by a whole number, write the decimal point in the quotient directly above the decimal point in the dividend. Then divide as with whole numbers.

Find: 9.92 ÷ 16

Write a decimal point in the quotient.	Divide.

$$16 \overline{)9.92}$$

$$\begin{array}{r} 0.62 \\ 16 \overline{)9.92} \\ \underline{96}\downarrow \\ 32 \\ \underline{32} \\ 0 \end{array}$$

Find: $48.96 ÷ 24

Write a decimal point in the quotient.	Divide.

$$24 \overline{)\$48.96}$$

$$\begin{array}{r} \$2.04 \\ 24 \overline{)\$48.96} \\ \underline{48}\downarrow\downarrow \\ 96 \\ \underline{96} \\ 0 \end{array}$$

Divide.

1.
$$\begin{array}{r} 8.2 \\ 8 \overline{)65.6} \\ \underline{64}\downarrow \\ 16 \\ \underline{16} \\ 0 \end{array}$$

$$5 \overline{)\$3.45}$$

$$3 \overline{)8.28}$$

$$7 \overline{)0.784}$$

2.
$$\begin{array}{r} 0.04 \\ 61 \overline{)2.44} \\ \underline{244} \\ 0 \end{array}$$

$$39 \overline{)\$58.50}$$

$$46 \overline{)9.338}$$

$$14 \overline{)\$43.96}$$

3. $7 \overline{)29.12}$

$4 \overline{)\$16.48}$

$71 \overline{)1.278}$

$22 \overline{)0.154}$

Set up the problem. Then divide.

4. 22.5 ÷ 15 = _____ $6.03 ÷ 9 = _____   114.8 ÷ 82 = _____

$$15 \overline{)22.5}$$

Dividing Decimals by Decimals

To divide a decimal by a decimal, change the divisor to a whole number by moving the decimal point. Move the decimal point in the dividend the same number of places. Then divide.

Remember, write a decimal point in the quotient directly above the new decimal point position in the dividend.

Find: 4.34 ÷ 0.7

Move each decimal point 1 place.	Divide.
0.7⤻)4.3⤻4	6.2 7)43.4 42↓ 14 14 0

Find: 0.0713 ÷ 0.23

Move each decimal point 2 places.	Divide.
0.23⤻)0.0⤻713	0.31 23)007.13 69↓ 23 23 0

Divide.

1.
$$1.7⤻)\overline{8.1⤻6}$$
4.8
68↓
136
136
0

0.5)2.6 4 5

4.6)0.0 1 3 8

3.9)$5 3.4 3

2.
$$0.1 6⤻)\overline{4.7 8⤻4}$$
29.9
32↓
158
144↓
144
144
0

0.2 4)1.4 8 8

0.0 8)$6.4 8

0.5 7)2.5 6 5

Set up the problem. Then divide.

3. 1.854 ÷ 0.9 = _____   0.91 ÷ 1.3 = _____ $15.18 ÷ 0.33 = _____

0.9)1.8 5 4

Dividing Smaller Numbers by Larger Numbers

Sometimes when you divide, the divisor will be larger than the dividend. To divide, add a decimal point and zeros as needed to the dividend. Continue to divide until the remainder is zero. In some cases, you may never have a remainder of zero. When dividing money, round the quotient to the nearest cent. Remember, zeros may be needed in the quotient also.

To change any fraction to its decimal equivalent, divide the denominator of the fraction into the numerator.

Find: $19 ÷ 300

Add a decimal point and zeros to the dividend.	Divide. Place a zero in the quotient.
$$300\overline{)\$19.00}$$	$$\begin{array}{r} \$0.0633 \\ 300\overline{)\$19.0000} \\ \underline{1800} \\ 1000 \\ \underline{900} \\ 1000 \\ \underline{900} \\ 100 \end{array}$$
Round $0.0633 to $0.06.	

Change $\frac{3}{4}$ to a decimal.

Divide the numerator by the denominator. Add a decimal point and zeros to the dividend.	Place a decimal point in the quotient. Divide until the remainder is zero.
$$4\overline{)3.00}$$	$$\begin{array}{r} 0.75 \\ 4\overline{)3.00} \\ \underline{28} \\ 20 \\ \underline{20} \\ 0 \end{array}$$
$\frac{3}{4} = 0.75$	

Divide. Write zeros as needed.

1. $$\begin{array}{r} 0.5 \\ 40\overline{)20.0} \\ \underline{200} \\ 0 \end{array}$$

$$100\overline{)\$30}$$

$$5\overline{)2}$$

$$20\overline{)\$15}$$

Change each fraction to its decimal equivalent.

2. $\frac{3}{8} = 0.375$

$$\begin{array}{r} 0.375 \\ 8\overline{)3.000} \\ \underline{24} \\ 60 \\ \underline{56} \\ 40 \\ \underline{40} \\ 0 \end{array}$$

$\frac{1}{4} =$ _____

$\frac{1}{5} =$ _____

$\frac{1}{2} =$ _____

3. $\frac{17}{200} =$ _____

$\frac{13}{25} =$ _____

$\frac{2}{5} =$ _____

$\frac{3}{40} =$ _____

Terminating and Repeating Decimals

The fractions $\frac{1}{4}$ and $\frac{3}{16}$ are examples of fractions that can be expressed as **terminating decimals.** Terminating decimals have a remainder of 0.

Many fractions cannot be expressed as terminating decimals because the division process never results in a remainder of zero. $\frac{1}{3}$ and $\frac{3}{11}$ are examples of fractions which *cannot* be expressed as terminating decimals. Note that no matter how far the division is carried out, there will always be a remainder. Also note that the quotients begin repeating themselves. These are examples of **nonterminating, repeating decimals.**

EXAMPLES

$$\frac{1}{4} = 4\overline{)1.00} \quad \begin{array}{r} 0.25 \\ \hline 8 \\ \hline 20 \\ 20 \\ \hline 0 \end{array}$$

$$\frac{3}{16} = 16\overline{)3.0000} \quad \begin{array}{r} 0.1875 \\ \hline 16 \\ \hline 140 \\ 128 \\ \hline 120 \\ 112 \\ \hline 80 \\ 80 \\ \hline 0 \end{array}$$

EXAMPLES

$$\frac{1}{3} = 3\overline{)1.000} \quad \begin{array}{r} 0.333 \\ \hline 9 \\ \hline 10 \\ 9 \\ \hline 10 \\ 9 \\ \hline 1 \end{array}$$

$$\frac{3}{11} = 11\overline{)3.0000} \quad \begin{array}{r} 0.2727 \\ \hline 22 \\ \hline 80 \\ 77 \\ \hline 30 \\ 22 \\ \hline 80 \\ 77 \\ \hline 3 \end{array}$$

Quotients may be rounded to the nearest tenth, hundredth, or thousandth. To round a decimal fraction:

(1) **Carry out the division to one more place than you need in your results.**
(2) **If the digit that is to be dropped is less than 5, drop it and make no other change.**
(3) **If the digit to be dropped is 5 or more, drop it and increase the digit to its left by 1.**

EXAMPLE

0.571
to nearest hundredth
0.57
to nearest tenth
0.6

Round each of the following decimals to the nearest thousandth.
($\approx$ means "is approximately equal to.")

1. $0.1684 \approx$ _____ $0.78672 \approx$ _____ $4.1643098 \approx$ _____ $6.2085 \approx$ _____

Divide and round quotients to the nearest hundredth.

2. $7\overline{)4}$ $99\overline{)67}$ $9.9\overline{)181}$ $330\overline{)107}$

Divide and round quotients to the nearest tenth.

3. $\frac{1}{3} =$ _____ $\frac{2}{3} =$ _____ $\frac{1}{6} =$ _____ $\frac{4}{9} =$ _____

Dividing by 10, 100, 1,000

To divide a decimal by a power of ten, move the decimal point in the dividend to the left as many places as there are zeros in the divisor.

Remember, sometimes you might need to write zeros in the quotient in order to correctly insert the decimal point.

Study these examples.

$0.89 \div 10 = 0.089$ $0.73 \div 100 = 0.0073$ $0.52 \div 1,000 = 0.00052$

$8.9 \div 10 = 0.89$ $7.3 \div 100 = 0.073$ $5.2 \div 1,000 = 0.0052$

Divide. Write zeros as needed.

1. $5 \div 100 = $ _____0.05_____ $0.5 \div 10 = $ _____ $0.6 \div 100 = $ _____

2. $0.50 \div 10 = $ _____ $0.55 \div 100 = $ _____ $0.61 \div 100 = $ _____

3. $1.25 \div 10 = $ _____ $12.5 \div 10 = $ _____ $0.0125 \div 100 = $ _____

4. $1.25 \div 100 = $ _____ $12.5 \div 100 = $ _____ $0.0125 \div 1,000 = $ _____

5. $1,265.15 \div 100 = $ _____ $347.85 \div 100 = $ _____ $0.36 \div 1,000 = $ _____

6. $16.5 \div 10 = $ _____ $1.65 \div 100 = $ _____ $0.036 \div 100 = $ _____

7. $14.92 \div 100 = $ _____ $275.3 \div 10 = $ _____ $0.15 \div 10 = $ _____

8. $14.92 \div 1,000 = $ _____ $275.3 \div 1,000 = $ _____ $0.15 \div 100 = $ _____

9. $642.15 \div 10 = $ _____ $642.15 \div 100 = $ _____ $64.251 \div 100 = $ _____

10. $0.375 \div 10 = $ _____ $3.75 \div 10 = $ _____ $37.5 \div 10 = $ _____

Solve.

▒▒▒▒▒▒▒▒ **Do Your Work Here** ▒▒▒▒▒▒▒▒

11. I paid $11.69 for 10 gallons of gas. How much is this per gallon?   _____

12. One box of oranges (100 to a box) sells for $15.00 per box. How much is this per orange?   _____

13. A farmer picked 150 bushels of cherries from 100 trees. What was the average picked from each tree?   _____

14. On the first day of her vacation, Mrs. Hausenfluke drove 195 miles. She used 10 gallons of gas. How many miles was this per gallon?   _____

15. A kilowatt means 1,000 watts, both being measures of electricity. How many kilowatts are there in 2,675 watts?   _____

Using Decimals and Fractions

Sometimes it may be easier to solve a problem using decimals, and at other times it is easier to solve a problem using fractions. Practice will tell you which method to use.

Find: $12.7 + 15\frac{1}{4}$

John lives 1.5 miles from town. Dan lives $1\frac{1}{4}$ miles from town. How far apart do John and Dan live?

Using fractions:

$12.7 = 12\frac{7}{10}$

$12\frac{7}{10} = 12\frac{14}{20}$

$+ 15\frac{1}{4} = 15\frac{5}{20}$

$27\frac{19}{20}$

Using decimals:

$15\frac{1}{4} = 15.25$

$\begin{array}{r} 15.25\ \checkmark \\ + 12.70 \\ \hline 27.95 \end{array}$

$1.5 = 1\frac{1}{2}$

$1\frac{1}{2} = 1\frac{2}{4}$

$- 1\frac{1}{4} = 1\frac{1}{4}$

$\frac{1}{4}$

$1\frac{1}{4} = 1.25$

$\begin{array}{r} 1.50\ \swarrow \\ - 1.25 \\ \hline 0.25 \end{array}$

John and Dan live $\frac{1}{4}$ or 0.25 mile apart.

Add or subtract using fractions. Then add or subtract using decimals.

1. $3.4 + 6\frac{1}{2} =$

$\begin{array}{r} 3\frac{4}{10} = 3\frac{4}{10} \\ + 6\frac{1}{2} = 6\frac{5}{10} \\ \hline 9\frac{9}{10} \end{array}$

$\begin{array}{r} 3.4 \\ + 6.5 \\ \hline 9.9 \end{array}$

$2.6 + 4\frac{1}{3} =$

$4.7 - 1\frac{3}{5} =$

2. $2.7 - 1\frac{1}{4} =$

$5.3 + 8\frac{7}{10} =$

$6.5 - 3\frac{3}{4} =$

3. $11\frac{1}{2} - 4.6 =$

$16\frac{3}{4} + 3.2 =$

$9\frac{1}{10} + 10.9 =$

Solve using both methods.

Do Your Work Here

4. A roadside sign says $8\frac{3}{4}$ miles to New Orleans. The odometer on our car measured the distance as 8.4 miles. How much difference was there between the sign and the odometer?

_____

5. The gauge (thickness) of a certain electric wire is marked 0.2 inch. A hole one-sixteenth inch wide has been drilled, and the wire is to be pushed through it. Will the wire fit the hole? (Hint: Compare 0.2 with $\frac{1}{16}$.)

_____

Choosing Decimals or Fractions

Read these word problems carefully. Work each problem using decimals or fractions.

1. You buy gasoline at 99.9 cents per gallon. This price is $7\frac{1}{4}$ cents more than the dealer pays. What is the price per gallon that the dealer pays?

   _____

2. A road sign read "Chicago $62\frac{1}{2}$ miles." The odometer of our car showed it to be 63.5 miles. How much difference was there in the two measurements?

   _____

3. The Malones have two cars. One car averages $27\frac{1}{2}$ miles to the gallon of gas. The other car averages 31.4 miles per gallon. How many miles more per gallon does the second car average?

   _____

4. Yesterday cotton sold for $33\frac{1}{2}$ cents a pound. Today it is selling for 33.8 cents a pound. How much is the increase?

   _____

5. The distance between two towns, according to the railroad timetable, was 13.7 miles. According to a sign along the road, it was $14\frac{1}{4}$ miles. How much difference was there between the two measurements?

   _____

6. The weather reporter said that yesterday's rain amounted to $2\frac{1}{2}$ inches. Today's rainfall, he said, was 1.8 inches. How much greater was the rainfall of yesterday?

   _____

7. By air it is 905.8 miles to Chicago. By road it is $1{,}109\frac{1}{2}$ miles. How many miles less is it by air?

   _____

8. A worker measured the gauge (width) of a piece of wire and said it was $\frac{1}{4}$ inch. His boss said it was actually 0.3 inch. How much was that difference?

   _____

9. Last year's rainfall amounted to $33\frac{1}{4}$ inches. Already this year there has been a rainfall of 17.6 inches. How much more rain must fall this year in order to equal the amount of last year?

   _____

Estimating Decimal Products and Quotients

To estimate decimal products or quotients, round each number to the nearest one. Then multiply or divide the rounded numbers.

Estimate: 15.3 × 2.8

Round each number.	Multiply.
15.3 → *15* 2.8 → *3*	$\begin{array}{r} 1\,5 \\ \times \quad 3 \\ \hline 4\,5 \end{array}$

Estimate: 360.41 ÷ 2.89

Round each number.	Divide.
360.41 → *360* 2.89 → *3*	$\begin{array}{r} 1\,2\,0 \\ 3\,\overline{)3\,6\,0} \\ \underline{3} \\ 0\,6 \\ \underline{6} \\ 0\,0 \\ \underline{0} \\ 0 \end{array}$

Estimate each product. Round to the nearest one.

1. $\begin{array}{r} 3\,1.7\,5 \rightarrow \\ \times \quad 2.2 \rightarrow \end{array}$ $\begin{array}{r} 3\,2 \\ \times \quad 2 \\ \hline 6\,4 \end{array}$ $\begin{array}{r} 1\,5.6 \rightarrow \\ \times \quad 3.5 \rightarrow \end{array}$ $\begin{array}{r} 2\,3.8 \rightarrow \\ \times \quad 4.7 \rightarrow \end{array}$

2. $\begin{array}{r} 9.5 \rightarrow \\ \times \,1\,3.4\,2 \rightarrow \end{array}$ $\begin{array}{r} 1.1\,7 \rightarrow \\ \times \quad 8.1 \rightarrow \end{array}$ $\begin{array}{r} 1\,2.6\,9 \rightarrow \\ \times \quad 6.2 \rightarrow \end{array}$

Estimate each quotient. Round to the nearest one.

3. $4.9\,\overline{)1\,9.9\,5} \rightarrow 5\,\overset{4}{\overline{)2\,0}}$ $6.2\,\overline{)1\,2.1\,7} \rightarrow$ $9.9\,5\,\overline{)1\,0\,9.5} \rightarrow$

4. $2\,2.1\,5\,\overline{)6\,5.7\,9} \rightarrow$ $3.6\,\overline{)2\,3.8} \rightarrow$ $2.9\,8\,\overline{)7\,4.9\,7} \rightarrow$

Estimate each answer. Round to the nearest one.

5. 45.8×9.3 $30.27 \div 10.14$ 31.996×4.2

$\begin{array}{r} 4\,6 \\ \times \quad 9 \end{array}$

99

Checking Up

Multiply.

1.
$$\begin{array}{r} 0.7 \\ \times\, 0.6 \\ \hline \end{array}$$
$$\begin{array}{r} 0.7 \\ \times\;\;\; 6 \\ \hline \end{array}$$
$$\begin{array}{r} 1 \\ \times\, 0.1 \\ \hline \end{array}$$
$$\begin{array}{r} 1.5 \\ \times\;\;\; 3 \\ \hline \end{array}$$
$$\begin{array}{r} 1.5 \\ \times\, 0.3 \\ \hline \end{array}$$
$$\begin{array}{r} 0.1\,5 \\ \times\;\; 0.3 \\ \hline \end{array}$$
$$\begin{array}{r} 3\,2 \\ \times\, 2.4 \\ \hline \end{array}$$
$$\begin{array}{r} 0.1\,2 \\ \times\, 0.1\,2 \\ \hline \end{array}$$

2.
$$\begin{array}{r} 1.2\,5 \\ \times\;\;\; 3\,0 \\ \hline \end{array}$$
$$\begin{array}{r} 0.0\,1\,5 \\ \times\;\;\; 0.1\,4 \\ \hline \end{array}$$
$$\begin{array}{r} 0.0\,7\,5 \\ \times\;\;\; 0.2\,2 \\ \hline \end{array}$$
$$\begin{array}{r} 1\,5.5 \\ \times\, 2.1\,2 \\ \hline \end{array}$$
$$\begin{array}{r} 7\,0\,5 \\ \times\, 2.0\,4 \\ \hline \end{array}$$

3. $2.25 \times 10 =$ _____ $0.225 \times 1{,}000 =$ _____ $22.5 \times 100 =$ _____

Change the fractions to decimals and multiply.

4. $0.056 \times 2\frac{1}{2} =$ $2.25 \times \frac{1}{4} =$ $18.9 \times 1\frac{1}{10} =$

_____ $\times$ _____ = _____   _____ $\times$ _____ = _____   _____ $\times$ _____ = _____

Divide. Round any repeating decimals to the nearest hundredth.

5. $4\overline{)2.4}$ $2\,4\overline{)0.9\,6}$ $2\,5\overline{)0.0\,7\,5}$ $0.4\overline{)2\,4}$ $0.1\,6\overline{)3\,2}$

6. $0.1\,8\overline{)9}$ $3.2\overline{)9\,6}$ $0.2\overline{)3.2}$ $0.1\,5\overline{)4.5}$ $0.2\,4\overline{)0.9\,6}$

7. $1.8\overline{)0.9}$ $1\,1\overline{)5}$ $8\overline{)7}$ $1\,6\overline{)5}$ $9\overline{)2}$

8. $35.2 \div 10 =$ _____ $18.6 \div 100 =$ _____ $36 \div 1{,}000 =$ _____

Change each fraction to its decimal equivalent. | Estimate each answer.

9. $\frac{4}{5} =$ _____ $\frac{1}{3} =$ _____   10. 56.8×11.3 $40.16 \div 7.95$

Solve the following.

1. A gallon of water weighs 8.355 pounds, and a gallon of milk weighs 1.03 times as much. How much does the milk weigh?   _____

2. A copper wire 48 feet long was divided into 15 pieces. How long was each piece?   _____

3. A kilowatt means 1,000 watts. How many watts are in 2,674 kilowatts?   _____

4. Ann Torres lives 6.4 miles from her work. She drives her car to and from work each day. How far does she drive in a 5-day week? (Hint: Think 10 trips per week.)   _____

5. Lee's property is 369.2 meters wide. If it is divided into 4 equal lots, how wide will each lot be?   _____

6. Beth cut her neighbor's lawn. It took 2.5 hours. She was paid $10.50. How much was this per hour?   _____

7. Dennis is making monthly car payments of $185.37. How much does he pay on the car in a year?   _____

8. A transport airplane flies the 810 miles from Hanover to Columbus in 3.75 hours. What average speed does it make?   _____

9. Nick bought 5 blades for his circular saw at $12.67 each. How much did he pay for the 5 blades?   _____

10. Maxine drove her car 201.25 miles on 11.5 gallons of gas. How many miles did she average to the gallon?   _____

11. Jack bought 3.5 pounds of apples at $0.72 per pound. How much did he pay?   _____

Write as decimals.

1. $\frac{1}{10}$ = _____ $\frac{5}{8}$ = _____ $1\frac{1}{5}$ = _____ $2\frac{3}{4}$ = _____

Write as a fraction or a mixed number.

2. 1.1 = _____   1.5 = _____   0.75 = _____   5.50 = _____

Find each answer.

3.
```
   7.7 6          1.9 2 3          0.0 9           0.0 1 8
   6.6 7          2.7 4 9          0.5 4           0.2 0 9
 + 4.3 9        + 1.6 3 7        + 0.9 7 8       + 4 0.0 9
```

4.
```
   7.9 4          0.5 0 6        3 0 6.0 9          7.4 7 6 3
 - 4.5 6        - 0.1 8 9 2      -  4 6.4 5       - 6.4 7 6 7
```

5.
```
   1.2 5            3.0 8            3.5              2.0 7 5
 × 0.0 4          ×   1 2          × 2.4            ×   0.0 8
```

6. 1 2)‾3.9 4 8   0.0 6)‾2 7.4 8   2 0)‾5   1 6)‾5

Estimate each answer. Round to the nearest one.

7. 36.04 + 8.17   27.16 − 8.35   23.4 × 6.8   47.72 ÷ 12.18

Multiply or divide.

8. 2.25 × 100 = _____   0.225 × 10 = _____   22.5 × 1,000 = _____

9. 35.2 ÷ 100 = _____   18.6 ÷ 10 = _____   149.2 × 100 = _____

Solve these problems.   ▒▒▒▒ Do Your Work Here ▒▒▒▒

10. The Schmidt family drove their car 296 miles on 16 gallons of gas. How many miles per gallon was this? _____

11. On another trip, the Schmidts drove 305.8 miles in 5.5 hours. What was the average speed? _____

Unit 4   Review

Making Sure of Whole Numbers—Addition

The next pages will show you how well you have learned the skills presented in this book.

Write the following number in words.

1. 10,769 is written _____ thousand, _____ hundred _____.

Add.

2.
$$
\begin{array}{r} 6 \\ 7 \\ 3 \\ +6 \\ \hline \end{array}
\quad
\begin{array}{r} 5 \\ 8 \\ 6 \\ +7 \\ \hline \end{array}
\quad
\begin{array}{r} 3 \\ 9 \\ 4 \\ +8 \\ \hline \end{array}
\quad
\begin{array}{r} 7 \\ 6 \\ 3 \\ +9 \\ \hline \end{array}
\quad
\begin{array}{r} 4 \\ 3 \\ 9 \\ +6 \\ \hline \end{array}
\quad
\begin{array}{r} 2 \\ 7 \\ 5 \\ +9 \\ \hline \end{array}
\quad
\begin{array}{r} 9 \\ 2 \\ 6 \\ +5 \\ \hline \end{array}
\quad
\begin{array}{r} 1 \\ 4 \\ 8 \\ +6 \\ \hline \end{array}
\quad
\begin{array}{r} 8 \\ 5 \\ 3 \\ +7 \\ \hline \end{array}
$$

3.
$$
\begin{array}{r} 28 \\ +57 \\ \hline \end{array}
\quad
\begin{array}{r} 56 \\ +68 \\ \hline \end{array}
\quad
\begin{array}{r} 62 \\ +92 \\ \hline \end{array}
\quad
\begin{array}{r} 74 \\ +18 \\ \hline \end{array}
\quad
\begin{array}{r} 81 \\ +25 \\ \hline \end{array}
\quad
\begin{array}{r} 44 \\ +65 \\ \hline \end{array}
\quad
\begin{array}{r} 93 \\ +87 \\ \hline \end{array}
\quad
\begin{array}{r} 14 \\ +76 \\ \hline \end{array}
$$

4.
$$
\begin{array}{r} 239 \\ +378 \\ \hline \end{array}
\quad
\begin{array}{r} 864 \\ +793 \\ \hline \end{array}
\quad
\begin{array}{r} 592 \\ +682 \\ \hline \end{array}
\quad
\begin{array}{r} 718 \\ +592 \\ \hline \end{array}
\quad
\begin{array}{r} 647 \\ +674 \\ \hline \end{array}
\quad
\begin{array}{r} 439 \\ +813 \\ \hline \end{array}
$$

5.
$$
\begin{array}{r} 5,613 \\ +1,879 \\ \hline \end{array}
\quad
\begin{array}{r} 7,683 \\ +5,809 \\ \hline \end{array}
\quad
\begin{array}{r} 6,357 \\ +3,775 \\ \hline \end{array}
\quad
\begin{array}{r} 9,084 \\ +1,684 \\ \hline \end{array}
\quad
\begin{array}{r} 306 \\ 7,659 \\ +6,497 \\ \hline \end{array}
\quad
\begin{array}{r} 4,501 \\ 4,623 \\ +5,713 \\ \hline \end{array}
$$

Estimate the sum. Round each number to the nearest hundred.

6.
$$
\begin{array}{r} 495 \rightarrow \\ +243 \rightarrow \\ \hline \end{array}
\qquad
\begin{array}{r} 749 \rightarrow \\ +350 \rightarrow \\ \hline \end{array}
\qquad
\begin{array}{r} 932 \rightarrow \\ +219 \rightarrow \\ \hline \end{array}
\qquad
\begin{array}{r} 1,465 \rightarrow \\ +\ 327 \rightarrow \\ \hline \end{array}
$$

Line up the digits. Then add.

7. 3,421 + 678 = _____   45 + 367 = _____   3,976 + 89 = _____

Solve.

8. A factory that makes electronic equipment produced 3,048 VCRs in June. In July, the factory produced 2,986 VCRs and another 2,809 in August. How many VCRs were produced in the three months together?

Answer _____

9. On Saturday, the Air and Water Show attracted 219,486 people. On Sunday, 321,908 people saw the show. How many people saw the Air and Water Show both days together? Round your answer to the nearest thousand.

Answer _____

Making Sure of Whole Numbers—Subtraction

Subtract. Check your answers by adding.

1.
```
  1 8        1 5        1 3        1 9        1 2        1 7        1 6        1 4
-   9      -   8      -   6      -   7      -   3      -   8      -   9      -   6
```

2.
```
  7 9        8 7        7 8        8 9        9 1        5 7        6 2        6 8
- 3 5      -   4      - 3 5      -   8      - 4 0      - 3 2      -   2      - 5 0
```

3.
```
  7 0        6 5        9 2        4 0        9 5        5 4        8 8        9 0
- 3 4      -   6      -   8      - 2 8      - 4 7      -   7      - 2 9      -   2
```

4.
```
  8 5 9      4 7 8      6 3 7      6 0 0      1 0 7      2 3 0      9 1 0
- 6 0 4    - 1 5 3    - 2 0 2    - 4 2 7    -   7 9    - 1 7 6    - 2 3 4
```

5.
```
  3,8 9 8      7,6 9 9      4,8 7 9      8,4 0 1      9 2,4 8 4      8 3,5 1 2
- 1,3 6 8    - 3,5 8 9    -   7 5 7    - 1,8 5 3    - 3 7,5 6 7    - 5 6,7 3 9
```

Estimate the difference. Round each number to the nearest hundred.

6.
```
    6 2 5 →          9 3 6 →          7 7 3 →          2,5 5 0 →
  - 3 5 7 →        - 1 9 2 →        - 2 4 7 →        -   7 9 5 →
```

Line up the digits. Then subtract.

7. 800 − 29 = _____   1,426 − 970 = _____   150 − 105 = _____

Solve.

8. It is 625 miles from Dallas to El Paso. It is 242 miles from Dallas to Houston. El Paso is how much farther from Dallas than Houston? _____

9. The average altitude in New Mexico is 5,700 feet. The average altitude in Minnesota is 1,200 feet. How much higher is the average altitude in New Mexico? Round your answer to the nearest thousand. _____

Multiply.

1. 4   6   3   9   6   5   8
   ×3   ×5   ×7   ×2   ×8   ×7   ×8

2. 4 1   6 2   8 3   2 0   7 2   5 0   4 2
   × 5   × 4   × 3   × 4   × 3   × 2   × 4

3. 5 3 6   4 0 3   4 9 2   8 0 1   7 1 0   6 4 9
   ×   7   ×   9   ×   8   ×   6   ×   8   ×   9

4. 2 8   2 5   3 8   2 8   9 0   2 0   1 8
   ×1 6   ×1 0   ×2 8   ×4 0   ×5 8   ×4 0   ×4 5

5. 3 1 0   3 5 4   7 6 4   1,9 3 6   2,4 9 7   3,8 0 5
   × 2 3   × 5 5   × 4 0   ×   3 7   ×   1 2   ×   2 6 9

Estimate the product. Round to the nearest ten.

6.   6 7 →   4 9 →   2 1 →   3 4 7 →
    ×4 2 →   ×7 9 →   ×1 9 →   × 2 8 →

Line up the digits. Then multiply.

7. 406 × 60 = ＿＿＿＿   3,990 × 42 = ＿＿＿＿   378 × 7 = ＿＿＿＿

Solve.

Do Your Work Here

8. A printing press can print 8,500 sheets of paper in one hour. How many sheets can be printed in 15 hours? ＿＿＿

9. A machine can fold 12,000 sheets of paper in one hour. How many sheets can be folded in 18 hours? ＿＿＿

Making Sure of Whole
Numbers—Division

Divide.

1. $9\overline{)72}$ $7\overline{)63}$ $8\overline{)64}$ $5\overline{)20}$ $4\overline{)28}$ $6\overline{)54}$ $3\overline{)18}$

2. $4\overline{)96}$ $6\overline{)84}$ $5\overline{)95}$ $8\overline{)96}$ $6\overline{)90}$ $7\overline{)84}$ $4\overline{)92}$

3. $7\overline{)637}$ $8\overline{)248}$ $9\overline{)639}$ $4\overline{)128}$ $5\overline{)155}$ $8\overline{)480}$ $5\overline{)150}$

4. $5\overline{)216}$ $2\overline{)119}$ $6\overline{)437}$ $8\overline{)378}$ $4\overline{)714}$ $5\overline{)424}$ $7\overline{)869}$

5. $45\overline{)315}$ $34\overline{)249}$ $56\overline{)280}$ $67\overline{)402}$ $72\overline{)586}$ $71\overline{)490}$

6. $47\overline{)1,709}$ $143\overline{)29,315}$ $34\overline{)2,008}$ $211\overline{)63,722}$ $123\overline{)13,340}$

Estimate the quotient. Round each number to the nearest ten.

7. $37\overline{)164} \rightarrow$ $25\overline{)449} \rightarrow$ $92\overline{)899} \rightarrow$

Solve.

8. Juanita used 3,240 screws to assemble 30 gym sets. How many screws did she use for each set? _____

9. Mr. Wong drove 250 miles on Monday, 275 miles on Tuesday, and 225 miles on Wednesday. How many miles did he average each day? _____

106

Change each fraction as indicated.

1. $\frac{1}{3} = \frac{}{12}$ $\frac{1}{2} = \frac{}{8}$ $\frac{1}{4} = \frac{}{16}$ $\frac{1}{4} = \frac{}{12}$ $\frac{1}{3} = \frac{}{9}$ $\frac{1}{2} = \frac{}{16}$

2. $\frac{2}{5} = \frac{}{10}$ $\frac{3}{4} = \frac{}{16}$ $\frac{4}{5} = \frac{}{15}$ $\frac{5}{6} = \frac{}{12}$ $\frac{5}{8} = \frac{}{16}$ $\frac{2}{3} = \frac{}{12}$

Reduce to lowest terms.

3. $\frac{9}{12} =$ $\frac{8}{12} =$ $\frac{9}{15} =$ $\frac{6}{12} =$ $\frac{6}{10} =$ $\frac{6}{9} =$

4. $\frac{15}{25} =$ $\frac{6}{8} =$ $\frac{12}{16} =$ $\frac{4}{16} =$ $\frac{2}{8} =$ $\frac{5}{15} =$

Reduce to whole numbers or mixed numbers.

5. $\frac{18}{15} =$ $\frac{14}{6} =$ $\frac{10}{4} =$ $\frac{9}{6} =$ $\frac{12}{8} =$ $\frac{13}{5} =$

6. $\frac{15}{8} =$ $\frac{7}{2} =$ $\frac{9}{4} =$ $\frac{15}{5} =$ $\frac{8}{2} =$ $\frac{9}{3} =$

Add. Reduce to lowest terms.

7.
$\begin{array}{r} \frac{1}{3} \\ + \frac{2}{3} \\ \hline \end{array}$
$\begin{array}{r} \frac{3}{5} \\ + \frac{2}{5} \\ \hline \end{array}$
$\begin{array}{r} \frac{5}{8} \\ + \frac{5}{8} \\ \hline \end{array}$
$\begin{array}{r} \frac{3}{8} \\ + \frac{5}{8} \\ \hline \end{array}$
$\begin{array}{r} \frac{1}{8} \\ + \frac{7}{8} \\ \hline \end{array}$
$\begin{array}{r} \frac{1}{4} \\ + \frac{3}{4} \\ \hline \end{array}$

8.
$\begin{array}{r} \frac{3}{9} \\ + \frac{1}{2} \\ \hline \end{array}$
$\begin{array}{r} \frac{1}{2} \\ + \frac{3}{8} \\ \hline \end{array}$
$\begin{array}{r} \frac{4}{7} \\ + \frac{1}{3} \\ \hline \end{array}$
$\begin{array}{r} \frac{1}{4} \\ + \frac{1}{2} \\ \hline \end{array}$
$\begin{array}{r} \frac{1}{5} \\ + \frac{3}{10} \\ \hline \end{array}$

9.
$\begin{array}{r} 9\frac{3}{4} \\ + \ \frac{3}{6} \\ \hline \end{array}$
$\begin{array}{r} 7\frac{5}{8} \\ + 6\frac{5}{6} \\ \hline \end{array}$
$\begin{array}{r} 15\frac{7}{8} \\ + \ 9\frac{3}{4} \\ \hline \end{array}$
$\begin{array}{r} \frac{1}{4} \\ + 8\frac{3}{4} \\ \hline \end{array}$

10.
$\begin{array}{r} 507 \\ + \ 19\frac{1}{2} \\ \hline \end{array}$
$\begin{array}{r} 109 \\ + \ 24\frac{7}{10} \\ \hline \end{array}$
$\begin{array}{r} 214 \\ + \ 15\frac{5}{6} \\ \hline \end{array}$
$\begin{array}{r} 125 \\ + \ 17\frac{3}{8} \\ \hline \end{array}$

Solve.

11. Ralph used $4\frac{1}{2}$ yards of solid blue fabric to make a banner. He used $2\frac{3}{4}$ yards of yellow fabric to trim the banner. How much fabric in all did he use on the banner?

Answer _____

12. In one day, Janet spent $2\frac{7}{8}$ hours varnishing an antique table. The next day she spent $3\frac{1}{2}$ hours varnishing a chair. How much time in all did she spend varnishing furniture?

Answer _____

Making Sure of Fractions—Subtraction

Change each mixed or whole number as indicated.

1. $12\frac{7}{10} = 11\frac{}{10}$ $8 = 7\frac{}{8}$ $4\frac{2}{5} = 3\frac{}{5}$ $9 = 8\frac{}{3}$ $7\frac{1}{4} = 6\frac{}{4}$

Subtract. Reduce to lowest terms.

2.

$$\frac{9}{10} - \frac{4}{10} \qquad \frac{2}{3} - \frac{1}{3} \qquad \frac{9}{16} - \frac{5}{16} \qquad \frac{7}{8} - \frac{3}{8} \qquad \frac{4}{5} - \frac{2}{5} \qquad \frac{3}{4} - \frac{1}{4}$$

3.

$$\frac{7}{8} - \frac{1}{3} \qquad \frac{1}{2} - \frac{1}{4} \qquad \frac{3}{5} - \frac{1}{4} \qquad \frac{7}{10} - \frac{3}{5}$$

4.

$$63\frac{3}{4} - \frac{1}{6} \qquad 22\frac{1}{2} - \frac{1}{5} \qquad 27\frac{5}{6} - \frac{1}{2} \qquad 18\frac{3}{4} - \frac{1}{4}$$

5.

$$9\frac{3}{4} - 2\frac{3}{10} \qquad 65\frac{7}{9} - 17\frac{3}{6} \qquad 204\frac{5}{8} - 35\frac{1}{6} \qquad 108\frac{7}{12} - 29\frac{1}{4}$$

6.

$$26 - 7\frac{2}{3} \qquad 14 - 9\frac{1}{4} \qquad 23 - 14\frac{1}{8} \qquad 6 - 2\frac{1}{8}$$

7.

$$21\frac{1}{3} - 12\frac{1}{2} \qquad 16\frac{2}{7} - 7\frac{1}{3} \qquad 32\frac{1}{3} - 14\frac{3}{4} \qquad 45\frac{1}{9} - 24\frac{1}{2}$$

Solve.

8. Betty ran 4 miles and Mark ran $2\frac{3}{4}$ miles. How much farther did Betty run? _____

9. Ho had $5\frac{1}{2}$ empty pages in his photo album. He filled $3\frac{1}{4}$ pages with pictures. How many pages did he have left to fill? _____

Making Sure of Fractions—Multiplication

Multiply. Simplify.

1. $\frac{2}{5} \times \frac{2}{3} =$ $\frac{5}{6} \times \frac{3}{4} =$ $\frac{2}{3} \times \frac{3}{7} =$

2. $\frac{3}{4} \times \frac{7}{12} =$ $\frac{1}{2} \times \frac{3}{8} =$ $\frac{12}{23} \times \frac{3}{4} =$

3. $\frac{15}{16} \times 4 =$ $20 \times \frac{2}{5} =$ $24 \times \frac{7}{10} =$

4. $\frac{1}{3} \times 4 =$ $\frac{1}{2} \times 3 =$ $3 \times \frac{1}{5} =$

5. $6\frac{1}{4} \times \frac{3}{5} =$ $\frac{3}{8} \times 4\frac{4}{5} =$ $1\frac{7}{8} \times \frac{4}{15} =$

6. $7\frac{2}{3} \times \frac{1}{2} =$ $2\frac{1}{2} \times \frac{1}{3} =$ $\frac{1}{2} \times 3\frac{1}{2} =$

7. $9\frac{1}{2} \times \frac{1}{8} =$ $2\frac{1}{3} \times \frac{6}{7} =$ $\frac{2}{5} \times 5\frac{1}{4} =$

8. $3\frac{1}{2} \times 1\frac{3}{5} =$ $2\frac{2}{9} \times 4\frac{1}{2} =$ $2\frac{9}{16} \times 2\frac{2}{3} =$

9. $5\frac{1}{3} \times 1\frac{3}{8} =$ $1\frac{1}{2} \times 1\frac{3}{4} =$ $4\frac{3}{5} \times 3\frac{1}{3} =$

Solve.

10. Peter kept $\frac{3}{5}$ of his salary for expenses. If $\frac{2}{3}$ of his expenses was spent on housing, what part of his salary was spent on housing?   _____

11. In a survey, $\frac{2}{3}$ of the people said they drove to work. Of those who drove to work, $\frac{1}{6}$ said they took the toll road. What part of those who drove to work took the toll road?   _____

Divide. Simplify.

1. $\frac{1}{4} \div \frac{1}{8} =$ $\frac{3}{5} \div \frac{1}{5} =$ $\frac{1}{3} \div \frac{1}{2} =$

2. $\frac{1}{2} \div \frac{1}{4} =$ $\frac{2}{9} \div \frac{7}{12} =$ $\frac{1}{3} \div \frac{3}{5} =$

3. $4 \div \frac{1}{2} =$ $8 \div \frac{4}{5} =$ $10 \div \frac{5}{6} =$

4. $12 \div \frac{3}{4} =$ $3 \div \frac{1}{2} =$ $\frac{1}{3} \div 9 =$

5. $5\frac{2}{3} \div 3 =$ $2\frac{1}{4} \div 6 =$ $8\frac{1}{2} \div 4 =$

6. $9\frac{7}{9} \div \frac{5}{6} =$ $6\frac{2}{3} \div \frac{1}{4} =$ $8\frac{2}{3} \div \frac{1}{3} =$

7. $5\frac{1}{2} \div \frac{1}{2} =$ $6\frac{2}{3} \div \frac{1}{5} =$ $1\frac{3}{8} \div \frac{2}{3} =$

8. $6\frac{4}{5} \div 1\frac{1}{5} =$ $12\frac{1}{3} \div 6\frac{3}{4} =$ $6\frac{2}{5} \div 5\frac{1}{3} =$

9. $7\frac{1}{2} \div 4\frac{3}{5} =$ $2\frac{1}{2} \div 1\frac{1}{3} =$ $2\frac{1}{4} \div 2\frac{1}{2} =$

Solve.   Do Your Work Here

10. Yolanda wants to make scarves from $\frac{3}{4}$ yard of silk. If she needs $\frac{1}{8}$ yard for each scarf, how many scarves can she make? _____

11. Raul ordered 7 pizzas for his party. Each person at the party was given $\frac{1}{3}$ of a pizza. If no pizza was left over, how many people attended Raul's party? _____

Making Sure of Decimals

Write these decimals in words.

1. 0.5 _____   7.1 _____   9.035 _____

Write each of these as decimals.

2. three tenths _____   fifty hundredths _____

3. fifteen ten-thousandths _____   fifteen thousandths _____

Write these decimal numbers, using the decimal point for "and."

4. five and five tenths _____   five and five hundredths _____

5. five and five thousandths _____   five and five ten-thousandths _____

6. fourteen hundred ninety-two and fourteen hundred ninety-two ten-thousandths _____

Change these fractions to decimals.

7. $\frac{1}{10}$ = _____ $\frac{1}{2}$ = _____ $\frac{1}{4}$ = _____ $\frac{3}{4}$ = _____ $\frac{3}{10}$ = _____ $\frac{7}{10}$ = _____

Change these decimals to fractions in lowest terms.

8. 0.1 = ____   0.5 = ____   0.50 = ____   0.8 = ____   0.75 = ____   0.25 = ____

Change these decimals to mixed numbers in lowest terms.

9. 1.5 = ____   2.25 = ____   3.75 = ____   4.1 = ____   7.50 = ____   9.3 = ____

Change these mixed numbers to decimals.

10. $1\frac{1}{2}$ = ____ $2\frac{1}{4}$ = ____ $3\frac{3}{4}$ = ____ $7\frac{1}{10}$ = ____ $4\frac{3}{10}$ = ____ $2\frac{2}{5}$ = ____

Draw a circle around the pairs that are equal.

11. 300 and 0.03   0.4 and $\frac{4}{10}$ 0.5 and .50   0.07 and $\frac{7}{10}$ 42 and 4.2

12. 150 and 0.150   210 and 2.10   35 and 30.5   0.25 and $\frac{1}{4}$ 0.5 and $\frac{1}{2}$

Draw a circle around the larger decimal in each pair.

13. 0.05 and 0.005   0.25 and 0.3   0.078 and 0.8   0.4 and 0.004   0.45 and 0.456

14. 1.5 and 1.15   4 and 0.45   0.017 and 1.7   2.5 and 2.25   3.08 and 3.1

Solve.   ░░░░░░░ **Do Your Work Here** ░░░░░░░

15. Henry earns $865.40 each week. How much does he earn in 16 weeks?   _____

16. Maureen earned $64.75 for 3.5 hours of work. How much does she earn per hour?   _____

Making Sure of Decimals

Add.

1.
```
   1 2 2.6        0.0 9        2 3.0 8 5       $3.0 0        4.3
       9.0 9      0.2 5            9.9          0.8 3       7.8 7
     1 9.7        0.8         1 0.0 1           0.9 4       4.9 0 8
 + 2 3 4.0 8 7  + 0.9 5 4    + 3 4 6.7      +   1.0 7    + 1 6.0 0 9
```

Subtract.

2.
```
   $1 6.9 2      3 5 7.5 3      2 4.0 5         2 5 0        1 4 3.9
 -     8.7 5   - 1 8 7.0 8    - 1 5.9        -   2 3.4    - 1 0 2
```

Multiply.

3.
```
    0.5 2        0.1 5 4        0.0 2 5          2 4        0.0 0 1 5
 ×      5      ×     0.2      ×     0.3        × 0.0 4    ×       0.0 3
```

4.
```
    4.0 9          2.5            2 0 5          4.0 8        4.1 2 5
 ×    2.2        × 3.2          ×   4.2        × 3 1 2      ×     0.0 8
```

Divide. Round quotients to the nearest hundredth when necessary.

5. $5\overline{)3}$ $6\overline{)7.2}$ $1 5\overline{)2.2 5}$ $2 5\overline{)3 7.5}$ $0.0 9\overline{)\$3.6 9}$

6. $0.7\overline{)4.9}$ $0.3 1\overline{)1 0.8 5}$ $1 0 0\overline{)2 5}$ $2 7\overline{)6 0}$ $0.0 1 5\overline{)0.0 2 2 5}$

Estimate by rounding to the nearest tenth.

7.
```
    4.3 2  →        3 0 0.4 5  →        7 7.9 1  →
 +  5.7 1  →     -   9 5.3 6  →      ×     3.8 9  →      9.8 )9 8 9.8 3 →
```

Change these fractions to decimals.

8. $\frac{11}{100} =$ _____ $\frac{10}{25} =$ _____ $\frac{3}{40} =$ _____ $\frac{21}{5} =$ _____ $1\frac{3}{4} =$ _____

Multiply.

9. $10 \times 12.5 =$ _____ $100 \times 1.25 =$ _____ $1,000 \times 0.125 =$ _____

Divide.

10. $12.5 \div 10 =$ _____ $1.25 \div 100 =$ _____ $1.25 \div 1,000 =$ _____

Basic Essentials of Mathematics

BOOK ONE

Mastery Test

Instructions

The test on pages 113-118 is designed to give you the opportunity to mark your progress at the end of this book. The teacher, of course, will be interested in seeing just how well you have done. But you will be most interested in seeing whether you have made the right amount of progress and in seeing how well you have retained all the skills studied to this point.

There is no time limit for this review. Use all the time necessary to complete it. If you come to a problem which you cannot work, go on to the others. Then when you have finished them, go back to any you have skipped. Do your best.

The review is divided into four parts: Whole Numbers, Fractions, Decimals, and Applying Your Skills. Each part is divided into the four operations: Addition, Subtraction, Multiplication, and Division. When you have finished the review and it has been scored, enter your scores in the chart on the right. This will enable you to determine exactly where there are any weaknesses and where you need to acquire more skill.

	PROBLEMS	PERFECT SCORE	MY SCORE
Whole Numbers			
Addition	1-7	7	_____
Subtraction	8-13	6	_____
Multiplication	14-19	6	_____
Division	20-25	6	_____
Fractions			
Addition	26-32	7	_____
Subtraction	33-38	6	_____
Multiplication	39-44	6	_____
Division	45-50	6	_____
Decimals			
Addition	51-57	7	_____
Subtraction	58-63	6	_____
Multiplication	64-69	6	_____
Division	70-75	6	_____
Applying Your Skills			
	76-100	25	_____
TOTAL SCORE		100	_____

Whole Numbers

Add.

1.
```
  9
  7
  5
  4
  6
  2
+ 8
```

2.
```
  1 2
  5 5
  7 4
  3 7
  4 9
+ 8 6
```

3.
```
  4 6 4
  6 4 7
  5 2 8
  7 8 6
  9 5 3
+ 2 7 5
```

4.
```
  2 8 5,3 4 6
+ 4 6 5,9 8 7
```

5.
```
  3 8 0,2 4 0
+   7 6,0 9 5
```

6.
```
  5,2 3 7
  4,7 8 5
  6,8 7 6
+ 8,5 6 9
```

7.
```
  4,1 0 5
  3,0 8 6
  2,4 7 0
+ 5,9 0 9
```

Subtract.

8.
```
  1 7
-   9
```

9.
```
  9 9
- 5 3
```

10.
```
  1,2 8 9
-   4 3 5
```

11.
```
  7 4 5
- 2 4 6
```

12.
```
  6 2 0
- 4 4 3
```

13.
```
  8 4,2 3 9
- 3 6,5 4 4
```

Multiply.

14.
```
  9
× 7
```

15.
```
  1 4
×   9
```

16.
```
  3 4
× 2 3
```

17.
```
  7 5
× 3 7
```

18.
```
  7 6 2
×   5 6
```

19.
```
  2 0 6
× 2 5 0
```

Divide.

20. $7\overline{)63}$

21. $7\overline{)287}$

22. $12\overline{)384}$

23. $9\overline{)2,007}$

24. $16\overline{)4,867}$

25. $126\overline{)27,095}$

114

Fractions

Add. Simplify.

26. $\frac{1}{8}$
$\frac{3}{8}$
$+ \frac{5}{8}$

27. $\frac{1}{2}$
$+ \frac{1}{4}$

28. $\frac{2}{3}$
$+ \frac{1}{4}$

29. $1\,5\frac{3}{4}$
$+\,1\,2\frac{1}{4}$

30. $1\,9\frac{7}{8}$
$+\,1\,2$

31. $2\,7\frac{1}{4}$
$+\,1\,3\frac{3}{8}$

32. $3\,0\frac{1}{2}$
$+\,1\,6\frac{1}{3}$

Subtract. Simplify.

33. $\frac{3}{4}$
$- \frac{1}{4}$

34. $\frac{3}{4}$
$- \frac{3}{8}$

35. $\frac{1}{3}$
$- \frac{1}{4}$

36. $2\,1\frac{3}{4}$
$-\quad 9$

37. $1\,9$
$-\quad 7\frac{2}{3}$

38. $3\,0\frac{1}{8}$
$-\,1\,4\frac{1}{4}$

Multiply. Simplify.

39. $5 \times \frac{3}{10} =$

40. $\frac{1}{2} \times \frac{4}{5} =$

41. $\frac{1}{2} \times 25 =$

42. $9 \times 2\frac{1}{3} =$

43. $8\frac{1}{3} \times \frac{3}{5} =$

44. $3\frac{1}{3} \times 2\frac{2}{5} =$

Divide. Simplify.

45. $1\frac{3}{8} \div 3 =$

46. $3 \div \frac{3}{8} =$

47. $\frac{1}{2} \div \frac{1}{6} =$

48. $4\frac{2}{3} \div \frac{7}{8} =$

49. $5\frac{1}{3} \div \frac{1}{2} =$

50. $10\frac{1}{2} \div 1\frac{3}{4} =$

Decimals

Add.

51.
```
   0.4
 + 0.5
```

52.
```
   0.2
   0.6
   0.3
 + 0.8
```

53.
```
   1 0 8.7 5
 + 2 3 5.8 7
```

54.
```
   2 1 4.0 8
 +   1 7.9
```

55.
```
   1.0 4 6
   0.8 7
   0.9
 + 4.1
```

56.
```
 $2 1 7.4 3
      7.1 5
    2 9.0 0
 +      4.3 5
```

57.
```
   2 6 3.0 0 0 5
 +     1 7.6
```

Subtract.

58.
```
  $0.9 0
 - 0.4 0
```

59.
```
   1 2 8.9
 -   1 9.8
```

60.
```
   2 1 6.5
 -     1 9
```

61.
```
   1 3 7
 -   1 5.8
```

62.
```
   2 5.0 0 4
 - 1 6
```

63.
```
   3 7 2
 - 1 3 5.0 0 6
```

Multiply.

64.
```
   0.4
 × 0.9
```

65.
```
   1.5
 × 0.9
```

66.
```
   0.0 2 5
 ×     0.3
```

67.
```
   0.1 5
 × 0.0 6
```

68.
```
   0.0 0 4
 ×     0.0 3
```

69.
```
   3.5 3
 × 2.4 0
```

Divide.

70. $0.5 \overline{)0.2\,5}$

71. $1\,5 \overline{)2.2\,5}$

72. $1.3 \overline{)0.0\,3\,9}$

73. $1.2 \overline{)4\,8}$

74. $0.1\,5 \overline{)7\,5}$

75. $5 \overline{)4}$

116

Applying Your Skills

Solve.

Do Your Work Here

76. Sam's truck weighs 4,725 pounds. The truck can carry 7,500 pounds. What is the total weight of the truck and full load? _____

77. One city has a population of about 11,566,740. Another city has a population of about 6,978,730. What would the combined population be? _____

78. The Cascade Tunnel in Washington is 41,152 feet long. The Moffat Tunnel in Colorado is 32,798 feet long. How much longer is the Cascade Tunnel? _____

79. Esther gave the storekeeper a twenty-dollar bill for a purchase of $14.51. How much change should she have received? _____

80. George made a payment of $37.50 on a bill of $73.94. How much did he still owe? _____

81. Bob Holstrom sold his farm of 87 acres for $700 an acre. How much did he receive for the farm? _____

82. A jet flew from New York to London in 7 hours. The average speed was 499.4 miles an hour. How many miles was the trip? _____

83. Sam can plant 261 cabbage plants in 1 row of his garden. How many rows in his garden must be used to plant 1,044 plants? _____

84. Last summer we took an auto trip of 3,427 miles. After we had driven 1,578 miles, how many miles were left? _____

85. For the first four months of the year, rain fell as follows: $2\frac{1}{2}$ inches, $3\frac{1}{4}$ inches, $1\frac{1}{8}$ inches, and $1\frac{1}{2}$ inches. What was the total rainfall for these four months? _____

86. Ron used $1\frac{3}{4}$ cups of sugar in making a cake and $\frac{3}{4}$ cup in making the icing. How much sugar did he use in all? _____

87. Last year Milwaukee had a total fall of snow amounting to 92 inches. The normal fall is $52\frac{1}{4}$ inches. How much above normal was the snowfall last year? _____

Applying Your Skills

88. The storekeeper sold $9\frac{1}{4}$ yards from a piece of material containing 53 yards. How much was left in the original piece? _____

89. Helen makes $66 a day. How much will she make if she works $5\frac{1}{2}$ days? _____

90. Last week we drove to Detroit. At an average speed of 50 miles per hour, the trip took $4\frac{1}{2}$ hours. How far was it? _____

91. Mr. Thomas shoveled $12\frac{1}{2}$ yards of sand in $2\frac{1}{2}$ hours. At that rate how many yards did he shovel in one hour? _____

92. How much material was required for each of two dresses of the same size if it took $5\frac{1}{2}$ yards for both? _____

93. Rain fell last month as follows: 2.4 inches, 0.19 inch, 3.7 inches, and 0.9 inch. What was the total rainfall? _____

94. Chicago to Memphis is 527.4 miles. Memphis to New Orleans is 393.8 miles. How far is Chicago from New Orleans? _____

95. The heaviest rainfall in one day on record for one city is 11.17 inches. The next heaviest is 8.30 inches. How much greater is the record fall? _____

96. Richard took a trip from Montreal to Vancouver, a distance of 2,937.67 miles. John went from Chicago to Denver, a distance of 1,014.9 miles. Richard went how much farther than John? _____

97. Since water weighs 8.355 pounds to the gallon, how much do 9.4 gallons weigh? _____

98. Milk weighs 8.605 pounds to the gallon. How much do 12.2 gallons weigh? _____

99. Mr. Lee drove his car 297.6 miles while using 16 gallons of gas. How many miles does he get to the gallon? _____

100. One of the new trains travels 217.5 miles in 3 hours. What average speed per hour does it make? _____

118

Answer Key

Unit 1

Page 5
1. 1 3 4 5 6 7 8 9 10
2. 3 7 4 11 9 10 2 8 6
3. 3 6 8 10 12 5 7 9 11
4. 6 4 10 12 5 7 9 11 13
5. 6 8 10 12 14 5 9 11 13
6. 7 9 8 10 11 13 12 6 15
7. 8 11 14 7 12 15 10 13 16
8. 8 13 17 10 14 11 15 16 12
9. 10 15 11 16 12 17 13 18 14

Page 6
1. one thousand, nine hundred fifty-four
2. one thousand, nine hundred seventeen
3. one thousand, eight hundred twelve
4. twenty-five thousand, four hundred sixteen
5. fourteen thousand, seven hundred three
6. ten thousand, nine hundred eight
7. twelve thousand, one hundred
8. twelve thousand, eight
9. ten thousand, ninety-two
10. one hundred twenty-three thousand, four hundred fifty-six
11. seven hundred fifty-six thousand, one hundred
12. one million, six hundred fifty-eight thousand, three hundred twenty-five

Page 7
1. 60 80 50 40 50
2. 740 4,770 3,640 5,560 50,000
3. 800 500 300 200
4. 68,500 56,600 24,100 79,500
5. 3,000 5,000 6,000 4,000
6. 18,000 44,000 37,000 12,000
7. 110 100
8. 1,780 1,800 2,000
9. 7,842,870 7,842,900 7,843,000 8,000,000

Page 8
1. 12 13 11 10 15 17 12 15 16
2. 19 15 13 14 13 15 13 19 18
3. 19 18 19 19 13 17 20 14 23
4. 17 6 18 19 19 23 22 21 23
5. 29 25 25 28 27 17 21 27 15
6. 14 19 18 15 14 15 17 12 12

Page 9
1. 97 94 79 79 99 79 91
2. 993 777 970 937 999 797 989
3. 7,879 6,799 7,689 8,888 9,699 7,379 8,977
4. 7,618 5,899 9,899 9,969 7,378 7,839 7,891
5. 4,796 9,194 8,899 8,829 7,987 7,889 6,997
6. 29,979 38,778 28,889 45,995 57,681 67,928

Page 10
1. 81 81 63 83 91 81
2. 764 571 882 954 863 871
3. 673 592 591 764 753 451
4. 910 621 810 821 614 810
5. 5,254 9,093 7,212 15,015 23,030
6. 801 381 5,534 10,590 8,126
7. 810 490 3,263

Page 11
1. 11 5 11 9 8 8 8 4 17
2. 29 28 28 27 17 29 26 27 29
3. 77 79 77 99 142 142 91 173
4. 1,189 788 999 1,323 1,372 601
5. 9,787 10,968 8,688 5,996 12,486 52,897
6. 11,406 11,304 12,501 11,148 16,905 81,984
7. 1,188 636
8. $71 9. $41,480

Page 12
1. 4 6 6 8 1 6 6 8
2. 5 0 9 8 4 3 7 3
3. 0 7 6 9 5 3 8 1
4. 6 7 9 1 5 6 8 0
5. 10 8 5 9 4 2 7 7
6. 7 4 5 7 2 9 8 3
7. 9 6 0 5 3 6 5 0
8. 7 6 9 3 8 1 7 7
9. 4 9 9 8 5 0 10 8

Page 13
1. 2 7 4 5 3 4 3 4 5
2. 7 9 5 9 9 8 5 9
3. 25 34 34 24 23 53 24 41
4. 29 44 32 36 58 45 47 13
5. 445 543 535 353 734 345
6. 742 554 944 653 753 733
7. 503 494 965 555 364 446

Page 14
1. 49 49 26 47 49 19 24 38
2. 382 483 296 276 309 388 479
3. 978 917 948 715 839 915 514
4. 648 733 545 725 943 874 813
5. 851 5 1,432 1,276 1,160 781 935

Page 15
1. 9 145 168 239 89 556 317
2. 398 478 387 97 544 198 746
3. 476 192 130 725 200 198 204
4. 43 192 278 30 2,465 1,297 1,536
5. 276 171 158

Page 16
1. 800 800 600 600
2. 1,100 800 700 900
3. 2,000 3,200 4,000 5,000
4. 300 300 200 100
5. 500 200 200 300
6. 3,100 2,600 1,000 6,100

Page 17
1. 3 3 1 8 9 5 7 8
2. 24 38 10 34 39 3 49 39
3. 511 211 291 253 379 277 192
4. 1,402 4,212 3,342 2,012 4,112 1,211
5. 2,554 5,266 1,214 984 1,450 201
6. 45,595 21,994 10,099 18,007 19,249 59,912
7. 5,601 337 670
8. 7,907 subscribers 9. 11,362 books

Page 18
1. addition
   2,089 miles
2. subtraction
   326 books
3. subtraction
   $187
4. addition
   314

Page 19
1. $550
2. 142,339 people
3. North Carolina
   18,636 square miles
4. 8,354 feet
5. 59,149 books
6. 4,250,000 miles
7. Arizona
   422,006 people
8. 1,012 cards
9. 9,867 people
10. 1,832 feet

Page 20
1. 24   14   25   15
2. 6   16   24   0
3. 45   28   48   63
4. 40   54   36   12
5. 42   56   30   9

Page 21
#								
1.	10	14	0	3	20	12	0	36
2.	40	15	0	24	16	0	8	45
3.	12	0	18	9	18	30	24	24
4.	7	0	54	9	40	21	0	32
5.	0	36	49	64	9	28	0	42
6.	81	63	0	27	8	32	48	0
7.	6	72	0	24	56	36	35	0
8.	0	45	27	7	20	42	63	54
9.	72	56	16	35	0	4	30	48

Page 22
#								
1.	48	66	68	60	77	48	90	63
2.	84	93	80	69	28	99	80	55
3.	144	156	460	1,089	882	610	528	968
4.	2,808	3,936	6,946	9,548	9,499	9,030		
5.	286,836	385,842	715,990	896,972	236,430			

Page 23
#						
1.	188	250	128	72	144	120
2.	320	186	360	128	142	129
3.	1,959	3,186	2,400	2,370	2,160	1,728
4.	2,568	2,643	2,240	2,680	1,888	1,448
5.	3,264	3,616	2,394	3,861	2,368	3,897
6.	2,428	1,832	3,800	1,518	2,020	3,400
7.	6,920	13,030	11,229	10,150	9,416	21,882

Page 24
#								
1.	817	1,512	2,275	1,296	742	2,001	1,440	1,188
2.	1,738	1,376	6,545	1,488	2,976	2,279	1,008	2,175
3.	5,735	6,700	16,280	5,698	8,740	26,316	43,148	
4.	55,012	59,856	90,093	128,968				
	77,392	113,520	219,375					
5.	7,595	21,720	800					

Page 25
#						
1.	90	120	70	240	240	400
2.	58,138	14,896	12,180	9,996	22,311	48,018
3.	6,250	3,650	31,850	4,960	27,420	31,160
4.	10,140	32,240	23,400	18,360	2,180	48,240
5.	42,000	6,000	30,000	20,000	21,000	27,000

Page 26
#							
1.	336	350	702	810	688	536	392
2.	5,760	0	4,950	7,136	3,627	3,752	
3.	5,495	849	6,030	4,675	4,760	2,592	
4.	1,325	1,840	5,220	741	1,120	1,850	250
5.	55,545	4,000	27,900	35,144	15,000	6,200	
6.	53,500	475,545	108,600	158,172	15,700	42,075	
7.	27,000	141,500	1,980				
8. 92,400 pounds   9. 3,500,000 pounds

Page 27
#							
1.	16	2	8				
2.	32	8	4				
3.	35	5	7				
4.	27	9	3				
5.	40	5	8				
6.	18	3	6				
7.	36	9	4				
8.	45	5	9				
9.	28	4	7				
10.	20	5	4				
11.	0	0	0				
12.	0	0	0				
13.	6	8	5	5	6	3	3
14.	8	7	7	5	2	7	8
15.	9	9	8	9	9	6	3
16.	5	1	2	4	5	3	6
17.	3	3	2	2	2	3	4

Page 28
#						
1.	2	3	4	5	6	7
2.	2	4	6	8	5	7
3.	2	5	8	4	6	9
4.	2	4	3	5	7	6
5.	3	6	5	9	4	7
6.	9	3	5	4	7	6
7.	3	6	2	4	8	5
8.	6	8	9	7	7	5
9.	7	9	7	9	9	6
10.	80	—	91	31	21	61
11.	20	21	51	51	51	62
12.	71	71	41	61	30	70

Page 29
#					
1.	23	16	16	13	12
2.	12	13	13	17	16

Page 30
#						
1.	16	12	23	17	13	14
2.	19	18	28	12	12	24
3.	58	64	43	33	32	32
4.	142	121	121	121	151	141

Page 31
#					
1.	23 R1	21 R3	11 R1	13 R1	11 R3
2.	141 R1	121 R3	114 R2	144 R2	124 R6
3.	29 R3	21 R1	31 R5	85 R3	78 R2

Page 32
1. too large   too small   too large
   1   4   4
2. too large   too large   too small
   3   2   3
3. too large   too large   too large
   7   2   1

Page 33
1. 213 R1   212   111 R10   221
2. 57 R24   535 R7   312   131 R8
3. 224   123 R408   123   224
4. 135 R9   341   163 R4   342

Page 34
1. 406   3,080   6,001
2. 3,034 R197   3,010 R5   2,041 R12
3. 203 R2   6,510   120 R113
4. 2,024 R19   405   3,009 R50

Page 35
1. 187
2. 165 pounds
3. 52 miles
4. 410 miles per hour
5. $40
6. 216 pounds

Page 36
1. 3,200   300   1,800   5,600
2. 2,400   4,800   2,000   2,700
3. 7,600   23,700   11,400   12,000
4. 20   20   29
5. 66   30   16
6. 6   20   13

Page 37
1. 12   24   15 R4   22   115   169
2. 31 R1   43 R1   25 R3   17 R6   62 R3   31 R7
3. 114 R1   111   211 R19   56   98 R7
4. 601 R6   2,000   106   406 R2
5. 900   6,000   6   4
6. 1,060 miles; averaged 106 miles per day
7. $5.00

Page 38
1. 26,880 oranges
2. 932 people
3. 133,300 miles
4. $7.00
5. 2,748 days
6. 12 hours
7. 145 crates
8. 1,250 bushels
9. $13,482
10. 12,552 miles
11. $102

Page 39
1. 1,313   4,037   20,000   7,879   8,192   25,855
2. 621   575   6,583   951   964   591
3. 868   16,828   4,384   9,702   94,163   727,584   25,000
4. 1 R3   19   342 R5   41 R2   234   704 R10
5. twelve thousand, three hundred thirteen
6. 37
7. 700   2,100   **8.** 1,600   9
9. 152 tapes   **10.** 432 tapes

Unit 2

Page 40
1. 4   **4.** $\frac{3}{4}$ **7.** $\frac{2}{3}$ $\frac{5}{8}$ $\frac{4}{6}$
2. 1   **5.** 4   **8.** $\frac{3}{2}$ $\frac{7}{6}$ $\frac{4}{4}$
3. $\frac{1}{4}$ **6.** 2   **9.** $\frac{3}{4}$ $\frac{5}{4}$ $\frac{1}{2}$

Page 41
1. improper, proper, improper
2. mixed number, proper, proper
3. proper, improper, mixed number
4. improper, proper, proper
5. mixed number, improper, improper

The student should have circled:
6. $\frac{6}{9}$, $\frac{5}{8}$
7. $\frac{1}{8}$
8. $\frac{6}{8}$, $\frac{9}{12}$, $\frac{1}{8}$, $\frac{20}{21}$, $\frac{1}{9}$
9. $\frac{3}{7}$, $\frac{1}{13}$
10. $\frac{5}{6}$, $\frac{3}{12}$

The student should have put an x on:
6. $\frac{12}{6}$, $\frac{13}{4}$, $\frac{8}{8}$
7. $\frac{4}{3}$, $\frac{9}{7}$, $\frac{5}{5}$
8. $\frac{18}{15}$, $\frac{6}{5}$
9. $\frac{15}{2}$, $\frac{6}{2}$, $\frac{16}{15}$
10. $\frac{9}{9}$, $\frac{4}{4}$, $\frac{10}{6}$, $\frac{12}{7}$

Page 42
1. $\frac{2}{3}$ is equivalent to $\frac{4}{6}$ $\frac{3}{5}$ is not equivalent to $\frac{3}{6}$
2. $\frac{3}{4}$ is equivalent to $\frac{6}{8}$ $\frac{3}{4}$ is not equivalent to $\frac{1}{4}$
3. $\frac{6}{8}$ is equivalent to $\frac{3}{4}$ $\frac{1}{2}$ is equivalent to $\frac{2}{4}$

Page 43
1. 6   6   3   2   10
2. 16   20   24   12   10
3. $\frac{2}{3}$ $\frac{1}{2}$ $\frac{3}{4}$ $\frac{1}{3}$ $\frac{3}{4}$
4. $\frac{1}{2}$ $\frac{2}{3}$ $\frac{1}{2}$ $\frac{4}{5}$ $\frac{1}{2}$
5. $3\frac{1}{2}$ $1\frac{1}{5}$ $3\frac{1}{4}$ $4\frac{1}{2}$ 3
6. $17\frac{1}{3}$ 4 $6\frac{1}{2}$ $33\frac{1}{3}$ $1\frac{1}{8}$

Page 44
1. $\frac{0}{5}$, $\frac{1}{5}$, $\frac{2}{5}$, $\frac{3}{5}$, $\frac{4}{5}$, $\frac{5}{5}$
2. $\frac{1}{8}$, $\frac{3}{8}$, $\frac{5}{8}$, $\frac{7}{8}$, $\frac{9}{8}$, $\frac{11}{8}$
3. $\frac{0}{4}$, $\frac{1}{4}$, $\frac{2}{4}$, $\frac{3}{4}$, $\frac{5}{4}$, $\frac{6}{4}$
4. $\frac{1}{6}$, $\frac{3}{6}$, $\frac{4}{6}$, $\frac{7}{6}$, $\frac{8}{6}$, $\frac{9}{6}$
5. $\frac{0}{11}$, $\frac{1}{11}$, $\frac{2}{11}$, $\frac{5}{11}$, $\frac{9}{11}$, $\frac{11}{11}$
6. $\frac{5}{16}$ $\frac{7}{16}$ $\frac{9}{16}$ $\frac{11}{16}$
7. $\frac{3}{4}$ $\frac{13}{16}$ $\frac{7}{8}$ $\frac{3}{8}$
8. $\frac{3}{8}$ $\frac{6}{8}$ $\frac{7}{8}$ $\frac{1}{6}$ $\frac{2}{6}$ $\frac{3}{6}$
9. $\frac{2}{8}$ $\frac{4}{8}$ $\frac{5}{8}$ $\frac{2}{8}$ $\frac{3}{8}$ $\frac{4}{8}$
10. $\frac{1}{8}$ $\frac{3}{8}$ $\frac{6}{8}$ $\frac{2}{6}$ $\frac{3}{6}$ $\frac{4}{6}$

Page 45
1. $\frac{5}{8}$ $\frac{3}{5}$ $\frac{4}{9}$ $\frac{3}{4}$ $\frac{7}{10}$ $\frac{5}{6}$ $\frac{2}{3}$ $\frac{5}{8}$
2. $\frac{11}{12}$ $\frac{9}{10}$ $\frac{11}{15}$ $\frac{9}{20}$ $\frac{8}{9}$ $\frac{5}{6}$ $\frac{7}{8}$ $\frac{7}{12}$
3. $\frac{3}{4}$ $\frac{2}{5}$ $\frac{5}{9}$ $\frac{7}{16}$ $\frac{1}{2}$ $\frac{2}{5}$ $\frac{1}{2}$ $\frac{1}{3}$
4. $2\frac{2}{3}$ $4\frac{1}{2}$ $1\frac{1}{4}$ $1\frac{1}{5}$ $1\frac{2}{3}$ $1\frac{1}{3}$ $1\frac{3}{10}$ $1\frac{1}{4}$
5. $1\frac{1}{4}$ $1\frac{3}{4}$ $1\frac{9}{10}$ $1\frac{7}{8}$ $2\frac{4}{5}$ $3\frac{1}{2}$ $3\frac{2}{3}$ 8

Page 46
1. $\frac{1}{3}$ $\frac{1}{2}$ $\frac{2}{3}$ $\frac{5}{8}$ $\frac{3}{8}$ $\frac{3}{4}$
2. $\frac{5}{12}$ $\frac{5}{8}$ $\frac{5}{6}$ $\frac{5}{6}$ $\frac{7}{8}$ $\frac{7}{8}$
3. $\frac{1}{2}$ $\frac{3}{5}$ $\frac{8}{9}$ $\frac{3}{4}$ $\frac{1}{2}$
4. $1\frac{1}{5}$ $1\frac{1}{6}$ $1\frac{3}{8}$ $1\frac{1}{4}$ $1\frac{1}{8}$
5. $1\frac{1}{12}$ $1\frac{3}{16}$ $1\frac{1}{2}$ $1\frac{1}{2}$ $1\frac{4}{15}$
6. $1\frac{1}{3}$ $1\frac{4}{9}$ $\frac{8}{9}$ $\frac{6}{7}$ $1\frac{2}{15}$

Page 47

1. $\frac{9}{12}$ $\frac{5}{15}$ $\frac{8}{20}$ $\frac{20}{24}$ $\frac{9}{12}$
   $\frac{2}{12}$ $\frac{12}{15}$ $\frac{5}{20}$ $\frac{21}{24}$ $\frac{4}{12}$
2. $1\frac{5}{18}$ $1\frac{7}{15}$ $\frac{19}{30}$ $\frac{13}{14}$ $1\frac{1}{9}$
3. $1\frac{5}{12}$ $1\frac{13}{30}$ $\frac{19}{28}$ $\frac{3}{4}$ $\frac{17}{20}$

Page 48

1. $1\frac{1}{4}$ $\frac{5}{8}$ 1 2 $1\frac{4}{5}$ $1\frac{1}{2}$
2. $2\frac{1}{3}$ $1\frac{4}{5}$ $1\frac{1}{2}$ $2\frac{1}{2}$ $1\frac{13}{16}$ $1\frac{2}{3}$
3. $2\frac{1}{5}$ $1\frac{3}{4}$ $2\frac{1}{15}$ $1\frac{1}{2}$ $1\frac{9}{16}$ $1\frac{2}{9}$
4. 4 2 3 2 4 3
5. $1\frac{11}{12}$ 3 $1\frac{5}{8}$ 2 $1\frac{5}{12}$ 4
6. $2\frac{1}{12}$ $2\frac{9}{10}$ $3\frac{1}{6}$ $4\frac{1}{5}$ $3\frac{1}{8}$ $5\frac{1}{4}$
7. $2\frac{2}{5}$ $2\frac{4}{7}$ $2\frac{7}{10}$ 2 $1\frac{5}{8}$ $2\frac{2}{5}$

Page 49

1. $6\frac{1}{6}$ $3\frac{1}{10}$ $4\frac{1}{14}$ $10\frac{1}{3}$ $9\frac{1}{4}$
2. $15\frac{7}{9}$ $5\frac{3}{8}$ $15\frac{7}{8}$ $9\frac{3}{5}$ $19\frac{5}{8}$
3. $9\frac{1}{14}$ $7\frac{1}{12}$ $10\frac{3}{10}$ $20\frac{3}{20}$ $20\frac{7}{24}$
4. $2\frac{3}{5}$ $4\frac{5}{12}$ $4\frac{13}{24}$ $22\frac{1}{10}$ $8\frac{7}{12}$
5. $37\frac{1}{2}$ $16\frac{11}{18}$ $41\frac{9}{10}$ $3\frac{13}{15}$ $30\frac{2}{9}$

Page 50

1. 4 4 4 3 3 8
2. 4 12 12 10 10 8
3. $\frac{3}{4}$ $\frac{2}{3}$ $\frac{3}{5}$ $\frac{1}{2}$ $\frac{3}{5}$ $\frac{2}{3}$
4. $\frac{3}{5}$ $\frac{3}{4}$ $\frac{3}{4}$ $\frac{1}{4}$ $\frac{1}{4}$ $\frac{1}{3}$
5. $1\frac{1}{5}$ $2\frac{1}{3}$ $2\frac{1}{2}$ $1\frac{1}{2}$ $1\frac{1}{2}$ $2\frac{3}{5}$
6. 3 4 3 $1\frac{7}{8}$ $3\frac{1}{2}$ $2\frac{1}{4}$
7. 1 1 $\frac{3}{4}$ $\frac{3}{4}$ 1
8. $1\frac{7}{15}$ $1\frac{1}{6}$ $\frac{7}{8}$ $\frac{8}{9}$ $1\frac{1}{12}$
9. $1\frac{1}{4}$ $1\frac{2}{5}$ $1\frac{1}{2}$ $1\frac{7}{8}$ $1\frac{9}{10}$
10. $10\frac{29}{30}$ $32\frac{5}{12}$ $21\frac{3}{8}$ $13\frac{5}{6}$ $21\frac{1}{10}$
11. $\frac{1}{4}$ in.

Page 51

1. $\frac{1}{10}$ $\frac{7}{16}$ $\frac{5}{12}$ $\frac{1}{9}$ $\frac{1}{4}$ $\frac{1}{6}$
2. $\frac{1}{2}$ $\frac{1}{2}$ $\frac{1}{3}$ $\frac{1}{3}$ $\frac{1}{2}$ $\frac{1}{4}$
3. $\frac{2}{3}$ $\frac{1}{3}$ $\frac{1}{3}$ $\frac{1}{2}$ $\frac{1}{4}$ $\frac{1}{2}$
4. $\frac{3}{16}$ $\frac{1}{4}$ $\frac{1}{2}$ $\frac{1}{5}$ $\frac{3}{5}$ $\frac{1}{3}$
5. $\frac{1}{2}$ $\frac{2}{3}$ $\frac{3}{4}$ $\frac{1}{5}$ $\frac{1}{5}$ $\frac{2}{5}$
6. $\frac{3}{5}$ $\frac{1}{3}$ $\frac{4}{5}$ $\frac{1}{4}$ $\frac{3}{8}$ $\frac{1}{2}$

Page 52

1. $\frac{3}{8}$ $\frac{1}{8}$ $\frac{1}{2}$ $\frac{1}{4}$ $\frac{1}{9}$
2. $\frac{1}{6}$ $\frac{1}{14}$ $\frac{3}{8}$ $\frac{13}{20}$ $\frac{1}{4}$
3. $\frac{7}{12}$ $\frac{1}{32}$ $\frac{1}{40}$ $\frac{1}{12}$ $\frac{1}{30}$
4. $\frac{7}{20}$ $\frac{17}{30}$ $\frac{7}{16}$ $\frac{7}{18}$ $\frac{11}{32}$
5. $3\frac{1}{6}$ $1\frac{5}{8}$ $2\frac{7}{15}$ $3\frac{1}{4}$ 1
6. $\frac{3}{5}$ $1\frac{11}{14}$ $1\frac{1}{8}$ $1\frac{23}{30}$ $\frac{5}{6}$

Page 53

1. $7\frac{3}{8}$ $9\frac{7}{24}$ $15\frac{1}{4}$ $12\frac{7}{12}$
2. $8\frac{1}{5}$ $16\frac{8}{21}$ $15\frac{1}{2}$ $19\frac{3}{10}$
3. $7\frac{1}{4}$ $16\frac{5}{12}$ $11\frac{2}{5}$ $7\frac{1}{14}$
4. $10\frac{3}{20}$ $6\frac{11}{30}$ $12\frac{5}{12}$ $8\frac{1}{16}$
5. $9\frac{7}{12}$ $6\frac{7}{15}$ $8\frac{1}{3}$ $9\frac{3}{10}$
6. $38\frac{7}{16}$ inches
7. $12\frac{3}{8}$ pounds

Page 54

1. $3\frac{1}{4}$ $\frac{2}{21}$ $1\frac{7}{12}$ $3\frac{2}{9}$
2. $6\frac{7}{15}$ $5\frac{1}{2}$ $3\frac{5}{12}$ $5\frac{1}{3}$
3. $18\frac{13}{24}$ $4\frac{4}{15}$ $6\frac{13}{30}$ $12\frac{3}{10}$
4. $8\frac{1}{5}$ $9\frac{3}{5}$ $5\frac{7}{30}$ $61\frac{1}{3}$
5. $3\frac{3}{8}$ $2\frac{3}{10}$ $1\frac{3}{14}$ $13\frac{4}{15}$
6. $3\frac{3}{16}$ $1\frac{5}{14}$ $5\frac{5}{24}$ $112\frac{1}{6}$

Page 55

1. 2 3 4 2 6
2. 7 9 8 10 15
3. $7\frac{3}{5}$ $4\frac{1}{4}$ $2\frac{1}{6}$ $1\frac{2}{3}$ $1\frac{5}{9}$
4. $7\frac{6}{11}$ $1\frac{1}{3}$ $7\frac{5}{7}$ $5\frac{1}{6}$ $6\frac{5}{12}$
5. $7\frac{2}{3}$ $1\frac{2}{5}$ $4\frac{1}{2}$ $2\frac{3}{4}$ $7\frac{1}{6}$
6. $\frac{1}{2}$ $9\frac{1}{4}$ $14\frac{8}{9}$ $\frac{10}{11}$ $19\frac{11}{15}$

Page 56

1. 7 6 9 13 3
2. $7\frac{7}{8}$ $2\frac{29}{30}$ $3\frac{3}{4}$
3. $8\frac{5}{6}$ $5\frac{5}{6}$ $7\frac{23}{30}$
4. $5\frac{13}{15}$ $8\frac{33}{35}$ $8\frac{7}{12}$
5. $7\frac{41}{42}$ $4\frac{7}{10}$ $7\frac{19}{20}$
6. $7\frac{17}{18}$ $4\frac{19}{20}$ $2\frac{7}{9}$

Page 57

1. $\frac{2}{3}$ $\frac{3}{4}$ $\frac{2}{3}$ $\frac{1}{2}$ $\frac{1}{2}$ $\frac{3}{8}$ $\frac{1}{2}$
2. 4 4 4 1 8 10
3. 5 3 11 4 6 7
4. $\frac{3}{4}$ $\frac{2}{3}$ $\frac{1}{2}$ $\frac{1}{4}$ $\frac{1}{4}$
5. $\frac{1}{12}$ $\frac{5}{8}$ $\frac{1}{6}$ $\frac{2}{9}$ $\frac{5}{18}$ $\frac{1}{4}$
6. $5\frac{1}{40}$ $6\frac{1}{8}$ $9\frac{4}{21}$ $12\frac{3}{10}$ $27\frac{1}{24}$
7. $8\frac{1}{4}$ $2\frac{1}{6}$ $3\frac{2}{5}$ $11\frac{5}{24}$
8. $9\frac{1}{2}$ $1\frac{2}{3}$ $7\frac{3}{5}$ $4\frac{1}{4}$ $10\frac{7}{12}$
9. $3\frac{5}{12}$ $6\frac{1}{2}$ $2\frac{3}{8}$ $10\frac{1}{3}$
10. $1\frac{1}{4}$ hours

Page 58

1. Multiply by 3.
   405, 1,215
2. Subtract 4.
   36, 32
3. Add $1\frac{1}{3}$.
   14
4. Divide by 4.
   80

Page 59

1. $1\frac{3}{8}$ cups
2. $\frac{3}{16}$ ounce
3. Bentley, $\frac{1}{2}$ dollar
4. $\frac{3}{8}$ inch
5. $1\frac{3}{8}$ cups
6. 24, 30
7. $4\frac{3}{4}$ pounds
8. $\frac{5}{8}$ mile
9. $1\frac{1}{4}$ pounds
10. 10

Page 60

1. $\frac{1}{9}$ $\frac{1}{12}$ $\frac{1}{4}$ $\frac{1}{6}$
2. $\frac{1}{10}$ $\frac{1}{10}$ $\frac{1}{6}$ $\frac{1}{25}$
3. $\frac{4}{27}$ $\frac{2}{5}$ $\frac{1}{2}$ $\frac{1}{4}$
4. $\frac{4}{15}$ $\frac{3}{25}$ $\frac{3}{20}$ $\frac{6}{25}$
5. $\frac{2}{9}$ $\frac{1}{5}$ $\frac{3}{50}$ $\frac{1}{18}$
6. $\frac{3}{16}$ $\frac{7}{50}$ $\frac{1}{16}$ $\frac{2}{7}$
7. $\frac{2}{25}$ $\frac{1}{5}$ $\frac{2}{49}$ $\frac{1}{9}$
8. $\frac{3}{5}$ $\frac{1}{10}$ $\frac{1}{7}$ $\frac{6}{35}$
9. 6 quarts
10. $\frac{3}{8}$ pound
11. $\frac{9}{32}$ mile

Page 61

1. $\frac{2}{7}$ $\frac{1}{15}$ $\frac{3}{10}$ $\frac{3}{4}$ $\frac{1}{14}$
2. $\frac{1}{4}$ $\frac{1}{10}$ $\frac{1}{7}$ $\frac{3}{8}$ $\frac{1}{2}$
3. $\frac{1}{6}$ $\frac{1}{4}$ $\frac{1}{9}$ $\frac{2}{9}$ $\frac{1}{6}$
4. $\frac{3}{5}$ $\frac{4}{7}$ $\frac{5}{14}$ $\frac{3}{22}$ $\frac{4}{9}$
5. $\frac{1}{3}$ $\frac{1}{4}$ $\frac{5}{18}$ $\frac{1}{3}$ $\frac{1}{2}$
6. $\frac{1}{3}$ $\frac{1}{9}$ $\frac{1}{8}$ $\frac{1}{3}$ $\frac{1}{6}$
7. $\frac{3}{8}$ mile 8. $\frac{3}{16}$ mile

Page 62

1. $1\frac{1}{3}$ $1\frac{1}{4}$ $1\frac{1}{5}$ 7. 15 30 15
2. $2\frac{2}{3}$ $2\frac{1}{2}$ $2\frac{2}{5}$ 8. 9 18 20
3. $3\frac{1}{2}$ $6\frac{3}{4}$ $4\frac{4}{5}$ 9. 6 $2\frac{6}{7}$ 15
4. 7 10 15   10. 21 12 $11\frac{1}{3}$
5. 8 4 10   11. 10 8 12
6. 1 5 50   12. $3\frac{3}{4}$ yards

Page 63

1. 6 $14\frac{1}{4}$ 26
2. 21 9 44
3. $7\frac{13}{15}$ $100\frac{1}{10}$ $43\frac{1}{2}$
4. 242 94 123
5. $14\frac{1}{4}$ $25\frac{3}{7}$ $7\frac{1}{2}$
6. $4\frac{9}{10}$ $16\frac{1}{10}$ 52
7. $19\frac{1}{4}$ $10\frac{1}{3}$ 51
8. $27\frac{1}{2}$ $5\frac{1}{2}$ $33\frac{1}{2}$

Page 64

1. $1\frac{3}{4}$ 2 $1\frac{1}{5}$ 6. $3\frac{5}{6}$ 2 1
2. $3\frac{1}{2}$ 1 $\frac{19}{24}$ 7. $\frac{5}{6}$ $1\frac{3}{16}$ $1\frac{1}{2}$
3. $\frac{4}{5}$ $1\frac{3}{4}$ $2\frac{5}{6}$ 8. 2 $2\frac{3}{16}$ $1\frac{3}{8}$
4. 3 $1\frac{1}{2}$ $2\frac{7}{12}$ 9. $\frac{3}{4}$ 1 $3\frac{1}{48}$
5. $1\frac{5}{16}$ $\frac{1}{2}$ $1\frac{1}{3}$

Page 65

1. $5\frac{13}{24}$ $3\frac{1}{3}$ $8\frac{1}{6}$ 6. $12\frac{2}{15}$ $6\frac{3}{4}$ $10\frac{23}{30}$
2. $10\frac{4}{5}$ $4\frac{19}{20}$ $7\frac{7}{8}$ 7. $2\frac{1}{5}$ $10\frac{5}{8}$ $10\frac{1}{45}$
3. $1\frac{7}{8}$ 10 $6\frac{7}{8}$ 8. $3\frac{1}{5}$ $10\frac{11}{15}$ $21\frac{2}{3}$
4. 10 $14\frac{7}{16}$ $1\frac{5}{16}$ 9. 15 8 $8\frac{47}{56}$
5. 9 $8\frac{2}{3}$ $6\frac{3}{4}$

Page 66

1. $4\frac{1}{2}$ $3\frac{2}{5}$ $3\frac{1}{3}$ $3\frac{1}{3}$ $2\frac{2}{3}$ $3\frac{3}{7}$
2. $2\frac{1}{2}$ $1\frac{1}{2}$ $5\frac{1}{3}$ $2\frac{2}{5}$ $2\frac{5}{8}$ $4\frac{3}{4}$
3. $\frac{9}{2}$ $\frac{5}{1}$ $\frac{19}{3}$ $\frac{29}{4}$ $\frac{3}{1}$ $\frac{33}{8}$
4. $\frac{27}{5}$ $\frac{31}{4}$ $\frac{9}{1}$ $\frac{33}{5}$ $\frac{153}{10}$ $\frac{101}{8}$
5. $\frac{3}{7}$ $\frac{3}{5}$ $\frac{1}{3}$ $\frac{5}{8}$ $\frac{25}{48}$
6. $\frac{1}{3}$ $\frac{2}{5}$ $\frac{1}{6}$ $\frac{3}{16}$ $\frac{1}{6}$
7. 9 5 3 4 9
8. 16 18 6 15 35
9. $2\frac{1}{6}$ 1 66 38 $1\frac{5}{18}$
10. 30 $\frac{5}{6}$ 25 $52\frac{1}{2}$ $81\frac{1}{4}$
11. $3.00 12. 8 yards 13. $7.00 14. Yes

Page 67

1. $\frac{1}{2}$ $\frac{1}{3}$ $\frac{1}{5}$ $\frac{1}{10}$
2. $\frac{2}{1}$ $\frac{3}{1}$ $\frac{4}{1}$ $\frac{5}{1}$
3. $\frac{3}{2}$ $\frac{4}{3}$ $\frac{8}{5}$ $\frac{8}{7}$
4. $\frac{4}{9}$ $\frac{1}{3}$ $\frac{1}{8}$ $\frac{1}{8}$ $\frac{1}{10}$
5. $\frac{2}{5}$ $\frac{1}{16}$ $\frac{5}{72}$ $\frac{7}{24}$ $\frac{5}{54}$
6. $\frac{3}{28}$ $\frac{5}{54}$ $\frac{5}{22}$ $\frac{7}{36}$ $\frac{9}{64}$

Page 68

1. 2 3 2 2 2
2. $1\frac{1}{5}$ 8 $\frac{1}{2}$ $\frac{5}{7}$ $\frac{5}{6}$
3. $\frac{8}{27}$ $\frac{5}{9}$ $1\frac{1}{2}$ $1\frac{1}{3}$ $2\frac{1}{10}$
4. 2 $\frac{1}{4}$ $\frac{5}{8}$ $\frac{2}{3}$ $\frac{7}{9}$
5. 24 squares
6. 48 shakers
7. 10 bars

Page 69

1. 25 16 27 16
2. 15 16 15 24
3. 36 24 27 64
4. $25\frac{1}{2}$ $24\frac{1}{2}$ 35 42
5. 35 80 50 48
6. 45 45 120 49
7. 81 70 100 128
8. 96 cans
9. 16 plots

Page 70

1. $\frac{1}{2}$ $\frac{1}{3}$ $\frac{1}{4}$ 6. $\frac{5}{16}$ $1\frac{1}{2}$ $\frac{1}{2}$
2. $1\frac{1}{2}$ $1\frac{3}{8}$ $1\frac{1}{3}$ 7. $4\frac{1}{2}$ $2\frac{1}{5}$ $\frac{3}{11}$
3. $\frac{2}{7}$ $2\frac{1}{2}$ $\frac{3}{8}$ 8. $1\frac{1}{3}$ $\frac{3}{8}$ $1\frac{1}{4}$
4. $1\frac{1}{2}$ $2\frac{1}{3}$ $1\frac{1}{2}$ 9. $\frac{7}{10}$ 2 $\frac{2}{3}$
5. $\frac{3}{10}$ $1\frac{1}{2}$ $1\frac{1}{5}$ 10. $2\frac{2}{3}$ $\frac{2}{9}$ $4\frac{1}{2}$

Page 71

1. 14 2 14
2. $2\frac{1}{9}$ $7\frac{3}{4}$ $7\frac{1}{3}$
3. 11 $7\frac{1}{2}$ $5\frac{1}{5}$
4. 2 $1\frac{1}{2}$ $7\frac{4}{5}$
5. $2\frac{3}{4}$ $2\frac{7}{9}$ $2\frac{3}{4}$
6. $2\frac{7}{10}$ 11 $11\frac{1}{2}$
7. $15\frac{3}{4}$ 4 $8\frac{1}{2}$
8. 6 9 $3\frac{2}{3}$
9. $3\frac{3}{32}$ $5\frac{4}{7}$ $14\frac{2}{5}$

Page 72

1. $\frac{1}{2}$ $1\frac{29}{41}$ $\frac{11}{18}$
2. $3\frac{2}{3}$ $1\frac{29}{46}$ $1\frac{2}{5}$
3. $3\frac{3}{10}$ $\frac{2}{3}$ $3\frac{5}{11}$
4. $3\frac{1}{33}$ $6\frac{1}{2}$ 4
5. 2 $5\frac{2}{3}$ $4\frac{2}{9}$
6. $3\frac{3}{5}$ $5\frac{1}{3}$ $2\frac{7}{10}$
7. $1\frac{7}{8}$ $\frac{3}{8}$ 3
8. $2\frac{26}{27}$ $2\frac{2}{3}$ $\frac{1}{2}$
9. $4\frac{2}{9}$ $1\frac{1}{3}$ $\frac{2}{3}$

Page 73

1. $\frac{13}{3}$ $\frac{5}{1}$ $\frac{32}{5}$ $\frac{8}{1}$ $\frac{47}{3}$ $\frac{103}{4}$ $\frac{17}{1}$ $\frac{201}{2}$
2. $5\frac{2}{3}$ $4\frac{3}{4}$ $6\frac{1}{5}$ $3\frac{1}{6}$ $2\frac{1}{8}$ $5\frac{2}{9}$ $3\frac{3}{10}$
3. $\frac{2}{3}$ $\frac{2}{3}$ $\frac{3}{4}$ $\frac{2}{5}$ $\frac{5}{9}$ $1\frac{1}{3}$ $5\frac{1}{5}$
4. $\frac{1}{6}$ $\frac{3}{2}$ $\frac{2}{1}$ $\frac{2}{9}$
5. $\frac{1}{16}$ $\frac{1}{10}$ $\frac{3}{25}$ $\frac{1}{8}$ $\frac{1}{6}$
6. 2 $\frac{1}{2}$ $\frac{2}{3}$ $\frac{5}{8}$ $\frac{5}{6}$
7. 12 15 25 $2\frac{1}{2}$ $4\frac{4}{5}$
8. $\frac{1}{3}$ $\frac{11}{20}$ $1\frac{3}{4}$ $\frac{1}{5}$ $1\frac{1}{2}$
9. $2\frac{1}{9}$ $7\frac{1}{2}$ 4 11 $2\frac{7}{9}$
10. $6\frac{2}{3}$ $1\frac{1}{5}$ $\frac{1}{2}$ $3\frac{3}{8}$ $\frac{3}{7}$
11. 4 fields   12. 8 stacks

Page 74

1. 12 gallons   5. 96 tubes
2. 4 frames   6. 15 bags
3. $\frac{5}{16}$ mile   7. $51
4. 7 sets   8. $3\frac{3}{4}$ yards

Page 75

1. $1\frac{17}{30}$ $9\frac{7}{20}$ $\frac{1}{2}$ $12\frac{13}{30}$ $3\frac{13}{24}$
2. $1\frac{3}{8}$ $6\frac{4}{15}$ $9\frac{13}{16}$ $6\frac{13}{20}$ $13\frac{8}{15}$
3. $1\frac{5}{12}$ $7\frac{3}{10}$ $13\frac{1}{12}$ $1\frac{3}{8}$ $10\frac{7}{9}$
4. $\frac{2}{3}$ $\frac{3}{5}$ $\frac{1}{2}$ 15
5. 16 $1\frac{3}{4}$ 15 $4\frac{1}{16}$
6. $\frac{1}{4}$ 10 $15\frac{5}{6}$ $3\frac{1}{16}$
7. $\frac{1}{16}$ $1\frac{1}{2}$ 51 $\frac{2}{3}$
8. 9 $\frac{1}{18}$ $1\frac{3}{10}$ 9
9. 16 5 $\frac{4}{5}$ $\frac{8}{9}$
10. $98\frac{1}{2}$ yards   11. 54 miles per hour
12. $3\frac{1}{6}$ dozen   13. 4 reams

124

0.1	0.625	1.2	2.75
$1\frac{1}{10}$	$1\frac{1}{2}$	$\frac{3}{4}$	$5\frac{1}{2}$
18.82	6.309	1.608	40.317
3.38	0.3168	259.64	0.9996
0.05	36.96	8.4	0.166
0.329	458	0.25	0.3125
44	19	161	4
225	2.25	22,500	
0.352	1.86	14,920	
18.5 mpg			
55.6 mph			

it 4

ge 103

ten thousand, seven hundred sixty-nine

22	26	24	25	22	23	22	19	23
85	124	154	92	106	109	180	90	
617	1,657	1,274	1,310	1,321	1,252			
7,492	13,492	10,132	10,768	14,462	14,837			
700	1,100	1,100	1,800					
4,099	412	4,065						

8,843 VCRs   **9.** 541,000 people

ge 104

9	7	7	12	9	9	7	8
44	83	43	81	51	25	60	18
36	59	84	12	48	47	59	88
255	325	435	173	28	54	676	
2,530	4,110	4,122	6,548	54,917	26,773		
200	700	600	1,800				
771	456	45					

383 miles   **9.** 5,000 feet

age 105

1. 12	30	21	18	48	35	64
2. 205	248	249	80	216	100	168
3. 3,752	3,627	3,936	4,806	5,680	5,841	
4. 448	250	1,064	1,120	5,220	800	810
5. 7,130	19,470	30,560	71,632	29,964	1,023,545	
6. 2,800	4,000	400	10,500			
7. 24,360	167,580	2,646				

8. 127,500 sheets   **9.** 216,000 sheets

age 106

1. 8	9	8	4	7	9	6
2. 24	14	19	12	15	12	23
3. 91	31	71	32	31	60	30
4. 43 R1	59 R1	72 R5	47 R2	178 R2	84 R4	124 R1
5. 7	7 R11	5	6	8 R10	6 R64	
6. 36 R17	205	59 R2	302	108 R56		
7. 4	15	10				

8. 108 screws   **9.** 250 miles

Page 107

1. 4	4	4	3	3	8
2. 4	12	12	10	10	8
3. $\frac{3}{4}$	$\frac{2}{3}$	$\frac{3}{5}$	$\frac{1}{2}$	$\frac{3}{5}$	$\frac{2}{3}$
4. $\frac{3}{5}$	$\frac{3}{4}$	$\frac{3}{4}$	$\frac{1}{4}$	$\frac{1}{4}$	$\frac{1}{3}$

5. $1\frac{1}{5}$	$2\frac{1}{3}$	$2\frac{1}{2}$	$1\frac{1}{2}$	$1\frac{1}{2}$	$2\frac{3}{5}$
6. $1\frac{7}{8}$	$3\frac{1}{2}$	$2\frac{1}{4}$	3	4	3
7. 1	1	$1\frac{1}{4}$	1	1	1
8. $\frac{5}{6}$	$\frac{7}{8}$	$\frac{19}{21}$	$\frac{3}{4}$	$\frac{1}{2}$	
9. $10\frac{1}{4}$	$14\frac{11}{24}$	$25\frac{5}{8}$	9		
10. $526\frac{1}{2}$	$133\frac{7}{10}$	$229\frac{5}{6}$	$142\frac{3}{8}$		

11. $7\frac{1}{4}$ yards   **12.** $6\frac{3}{8}$ hours

Page 108

1. 17	8	7	3	5	
2. $\frac{1}{2}$	$\frac{1}{3}$	$\frac{1}{4}$	$\frac{1}{2}$	$\frac{2}{5}$	$\frac{1}{2}$
3. $\frac{13}{24}$	$\frac{1}{4}$	$\frac{7}{20}$	$\frac{1}{10}$		
4. $63\frac{7}{12}$	$22\frac{3}{10}$	$27\frac{1}{3}$	$18\frac{1}{2}$		
5. $7\frac{9}{20}$	$48\frac{5}{18}$	$169\frac{11}{24}$	$79\frac{1}{3}$		
6. $18\frac{1}{3}$	$4\frac{3}{4}$	$8\frac{7}{8}$	$3\frac{7}{8}$		
7. $8\frac{5}{6}$	$8\frac{20}{21}$	$17\frac{7}{12}$	$20\frac{11}{18}$		

8. $1\frac{1}{4}$ miles   **9.** $2\frac{1}{4}$ pages

Page 109

1. $\frac{4}{15}$	$\frac{5}{8}$	$\frac{2}{7}$
2. $\frac{7}{16}$	$\frac{3}{16}$	$\frac{9}{23}$
3. $3\frac{3}{4}$	8	$16\frac{4}{5}$
4. $1\frac{1}{3}$	$1\frac{1}{2}$	$\frac{3}{5}$
5. $3\frac{3}{4}$	$1\frac{4}{5}$	$\frac{1}{2}$
6. $3\frac{5}{6}$	$\frac{5}{6}$	$1\frac{3}{4}$
7. $1\frac{3}{16}$	2	$2\frac{1}{10}$
8. $5\frac{3}{5}$	10	$6\frac{5}{6}$
9. $7\frac{1}{3}$	$2\frac{5}{8}$	$15\frac{1}{3}$

10. $\frac{2}{5}$

11. $\frac{1}{9}$

Page 110

1. 2	3	$\frac{2}{3}$
2. 2	$\frac{8}{21}$	$\frac{5}{9}$
3. 8	10	12
4. 16	6	$\frac{1}{27}$
5. $1\frac{8}{9}$	$\frac{3}{8}$	$2\frac{1}{8}$
6. $11\frac{11}{15}$	$26\frac{2}{3}$	26
7. 11	$33\frac{1}{3}$	$2\frac{1}{16}$
8. $5\frac{2}{3}$	$1\frac{67}{81}$	$1\frac{1}{5}$
9. $1\frac{29}{46}$	$1\frac{7}{8}$	$\frac{9}{10}$

10. 6 scarves

11. 21 people

1991 Steck-Vaughn Company.

Page 111
1. five tenths; seven and one tenth; nine and thirty-five thousandths
2. 0.3   0.50
3. 0.0015   0.015
4. 5.5   5.05
5. 5.005   5.0005
6. 1,492.1492
7. 0.1   0.5   0.25   0.75   0.3   0.7
8. $\frac{1}{10}$ $\frac{1}{2}$ $\frac{1}{2}$ $\frac{4}{5}$ $\frac{3}{4}$ $\frac{1}{4}$
9. $1\frac{1}{2}$ $2\frac{1}{4}$ $3\frac{3}{4}$ $4\frac{1}{10}$ $7\frac{1}{2}$ $9\frac{3}{10}$
10. 1.5   2.25   3.75   7.1   4.3   2.4
11. Circle 0.4 and $\frac{4}{10}$, 0.5 and 0.50.
12. Circle 0.25 and $\frac{1}{4}$, 0.5 and $\frac{1}{2}$.
13. Circle 0.05, 0.3, 0.8, 0.4, 0.456.
14. Circle 1.5, 4, 1.7, 2.5, 3.1.
15. $13,846.40
16. $18.50

Page 112
1. 385.477   2.094   389.695 $5.84   33.087
2. $8.17   170.45   8.15   226.6   41.9
3. 2.60   0.0308   0.0075   0.96   0.000045
4. 8.998   8.00   861.0   1,272.96   0.33000
5. 0.6   1.2   0.15   1.5 $41.00
6. 7   35   0.25   2.22   1.5
7. 10   205.1   303.81   101
8. 0.11   0.4   0.075   4.2   1.75
9. 125   125   125
10. 1.25   0.0125   0.00125

Mastery Test

Page 114
1. 41
2. 313
3. 3,653
4. 751,333
5. 456,335
6. 25,467
7. 15,570
8. 8
9. 46
10. 854
11. 499
12. 177
13. 47,695
14. 63
15. 126
16. 782
17. 2,775
18. 42,672
19. 51,500
20. 9
21. 41
22. 32
23. 223
24. 304 R3
25. 215 R5

Page 115
26. $1\frac{1}{8}$
27. $\frac{3}{4}$
28. $\frac{11}{12}$
29. 28
30. $31\frac{7}{8}$
31. $40\frac{5}{8}$
32. $46\frac{5}{6}$
33. $\frac{1}{2}$
34. $\frac{3}{8}$
35. $\frac{1}{12}$
36. $12\frac{3}{4}$
37. $11\frac{1}{3}$
38. $15\frac{7}{8}$
39. $1\frac{1}{2}$
40. $\frac{2}{5}$
41. $12\frac{1}{2}$
42. 21
43. 5
44. 8
45. $\frac{11}{24}$
46. 8
47. 3
48. $5\frac{1}{3}$
49. $10\frac{2}{3}$
50. 6

Page 116
51. 0.9
52. 1.9
53. 344.62
54. 231.98
55. 6.916
56. $257.93
57. 280.6005
58. $0.50
59. 109.1
60. 197.5
61. 121.2
62. 9.004
63. 236.994
64. 0.36
65. 1.35
66. 0.0075
67. 0.0090
68. 0.00012
69. 6.4720
70. 0.5
71. 0.15
72. 0.03
73. 40
74. 500
75. 0.8

Page 117
76. 12,225 pounds
77. 18,545,470 people
78. 8,354 feet
79. $5.49
80. $36.44
81. $60,900
82. 3,495.8 miles
83. 4 rows
84. 1,849 miles
85. $8\frac{3}{8}$ inches
86. $2\frac{1}{2}$ cups
87. $39\frac{3}{4}$ inches

Page 118
88. $43\frac{3}{4}$ yards
89. $363
90. 225 miles
91. 5 yards
92. $2\frac{3}{4}$ yards
93. 7.19 inches
94. 921.2 miles
95. 2.87 inches
96. 1,922.77 miles
97. 78.537 pounds
98. 104.981 pounds
99. 18.6 miles per gallon
100. 72.5 miles per hour